the Tomato book

the Tomato book

GAIL HARLAND
SOFIA LARRINUA-CRAXTON

London, New York, Melbourne,
Munich, and Delhi

Editor Andrew Roff
Designer Kathryn Wilding
Senior Art Editor Susan Downing
Managing Editor Dawn Henderson
Managing Art Editor Christine Keilty
Senior Jacket Creative Nicola Powling
Production Editor Ben Marcus
Production Controller Wendy Penn
Creative Technical Support Sonia Charbonnier

Photography
Sarah Ashun Graham Rae
William Reavell Howard Rice

DK India
Head of Publishing Aparna Sharma
Art Director Shefali Upadhyay
Designer Neha Ahuja
DTP Designer Tarun Sharma
Production Manager Pankaj Sharma

First published in Great Britain in 2009
by Dorling Kindersley Limited
80 Strand, London WC2R 0RL

Penguin Group (UK)

Copyright © 2009 Dorling Kindersley Limited
Text copyright © 2009 Dorling Kindersley Limited
Recipe text copyright © 2009 Sofia Larrinua-Craxton

2 4 6 8 10 9 7 5 3 1

A CIP catalogue record for this book
is available from the British Library

ISBN 978-1-4053-4118-9

Colour reproduction by MDP, UK

Printed and bound in China
by Hung Hung Printing

**Discover more at
www.dk.com**

Contents

The tomato story

Tangy, acidic, sweet, and juicy – the tomato's flavour profile puts it in prime position in kitchens around the world. But there is more to a tomato than its taste ...

Discovering the "golden apple"

Tomatoes originated in the coastal highlands of western South America where they were grown by the Aztecs and Mayans. They grow wild in Ecuador, northern Chile, Peru, and the Galapagos Islands – they are thought to have been brought here in the stomachs of turtles. The first written description of tomatoes, by the Italian Pietro Andrea Matthiola in 1544, referred to them as *Mala aurea*, or "golden apples", however their close relationship to poisonous plants, such as the woody nightshade vine, meant they didn't receive the same regard when introduced more widely. (This suspicion isn't completely unfounded – tomato leaves do contain the poisonous chemical tomatine.) The tomato's great flavour didn't hide for long, however, and it soon became a welcome addition to cuisines around the world.

A wealth of varieties

New varieties arose by natural cross-pollination and selection, bringing us the largest beefsteaks to the smallest cherry tomatoes. Now, with over 5,000 varieties of all shapes and sizes, you will be able to find ones that suit your taste and planting space – be it a sunny windowsill, pot, plot, or patio – perfectly. Healthy, easy to grow, and easy to care for, it's no wonder they are the most popular home crop. Use the following pages to choose which varieties to grow, then find out how best to nurture them, and finally choose a recipe that will allow you to enjoy your harvest in all its glory ...

Exotic Fruiting tomato plants are a wonderful
sight in the garden; no wonder they are
sometimes referred to as "love apples".

The tomatoes

Choose from the following tomato varieties to decide what to grow, nurture, and enjoy. Think about where you will be growing your tomatoes, how much space you can give them, and in what types of recipes you wish to enjoy them. Please note, the tomatoes in the following pages are categorized according to their shape and size and categories may differ in other sources.

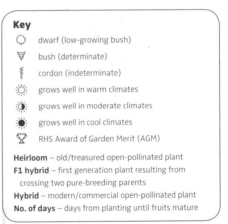

Key

◯	dwarf (low-growing bush)
⩒	bush (determinate)
ⵙ	cordon (indeterminate)
☼	grows well in warm climates
◑	grows well in moderate climates
☀	grows well in cool climates
♔	RHS Award of Garden Merit (AGM)

Heirloom – old/treasured open-pollinated plant
F1 hybrid – first generation plant resulting from crossing two pure-breeding parents
Hybrid – modern/commercial open-pollinated plant
No. of days – days from planting until fruits mature

The plant

The tomato plant is a short-lived annual that thrives in temperate regions. Plants grow very differently, so don't worry about not having enough space – there is a plant for every sized pot (or plot!)

Tomato plants range in height from tiny plants just 15cm (6in) tall to vines that can reach 10m (30ft), although 2.5m (8ft) is usual. There are three main growth habits: dwarf, bush, and cordon.

Dwarf types, such as 'Tiny Tim' (p37) and 'Micro-Tom' (p43), have a dwarfing gene, which makes these compact plants ideal for growing in containers. Many dwarf plants look very ornamental when in full bloom.

Bush types, such as 'Siberian' (p34) and 'Roma' (p65), are called determinates, as they usually grow to a fixed height. A bush has no prominent leading shoot. Instead it has a number of sideshoots, each ending in a fruit truss, which form a bushy, often sprawling plant. It usually fruits in one flush and so is useful for commercial use, as it is easy to harvest altogether.

Cordon, or vine plants, such as 'Jubilee' (p22) and 'Extra Sweetie' (p46), are indeterminates, as their growth height is not fixed. These vigorous plants have one stem that continues to flower and fruit, on trusses from the main stem, over a long season until it is killed by frost.

A few plants have variable habits and are called semi-bush, or semi-determinate, types. They usually reach a set height like bush tomatoes, but produce a second crop.

Heirlooms, F1 hybrids, and hybrids

Tomato plants are often classed as either heirloom or hybrid varieties. There are many definitions of these terms with some people maintaining that heirloom plants must have been grown for at least 50 years (or even 100 years). Others argue that true heirlooms are those that have been passed down through generations of a single family. Many so-called heirlooms are recent creations but have attracted the title, as they are unlikely to be commercially successful and so their survival depends on gardeners saving their seed. In this book, the term heirloom is used in its widest sense to refer to any open-pollinated variety that is particularly treasured or has a known heritage.

F1 hybrid plants are the first generation that result from crossing two selected pure-breeding parents. They are commercially produced and seed saved from these plants will not breed true so the seedlings may vary and be of a lower quality.

There are also a number of modern or commercial open-pollinated varieties here called hybrids. They were originally produced by cross-breeding genetically distinct parents but have been stabilized so that open-pollinated seed will usually produce offspring like the parent plant.

GROWTH HABITS – choose a type to suit the way you wish to grow it

Dwarf Most dwarf plants grow to a very small size and are ideal for growing on a patio or windowsill.

Bush These short and shapely plants have many sideshoots, which mean the plant sprawls out in all directions.

Cordon These long and upright plants produce a range of fruit shapes and colours. Their fruiting season is longer too.

The leaves

Most gardeners do not consider the foliage of the tomato plant, but the leaves of some are quite distinctive and can cause the plant to have an increased resistance to disease.

The regular tomato leaf is composed of a number of leaflets, each of which has a serrated, irregular edge, or margin.

Potato-leafed plants, such as 'Hillbilly Potato Leaf' (p53), may have a few lobes on their leaves, but generally the edges are smooth. They are slightly thicker than those of regular tomato plants and they therefore may be more resistant to some diseases.

Some plants, such as 'Green Sausage' (p77), have more finely divided leaves than others; the Russian heirloom 'Silvery Fir Tree' (p28) is particularly known for its very fine leaves. The plant 'Variegated', thought to have originated in Ireland, has leaves irregularly splashed with cream. Most attractive for their foliage are plants such as 'Elberta Girl' and 'Velvet Red' (p43); these have leaves so densely covered in silvery hairs that they are quite furry to the touch. This type of leaf is sometimes called an angora leaf, after the tomato variety of that name. It is important to keep an eye on the foliage of your tomato plants as they grow because the first signs of many problems can occur there (pp114–17).

LEAF TYPES – don't reach straight for the fruit, the leaves like to show off too!

Regular Most plants, such as 'Ildi', have leaves with a toothed margin.

Potato A leaf form with smooth edges that may be more resistant to disease.

Serrated Some plants have quite fine leaves, such as 'Green Sausage'.

Angora Covered in fine silvery hairs, these plants repel insect attack.

The fruits

Tomato fruits vary in size from currant tomatoes that weigh less than 5g (¼oz) to monster beefsteaks of 1kg (2lb) or more. They can also be found in a rainbow of colours from creamy white to deep purple.

Shapes and sizes

Tomatoes are usually classified according to their shape and size. The familiar round tomato is the standard globe. Small round tomatoes are known as cherries and those slightly bigger, such as 'Garcia', usually marketed on the vine, are called cocktail tomatoes. Very tiny fruits are referred to as currant tomatoes. Beefsteak tomatoes can sometimes be the size of a small pumpkin and have a flattened globe-like appearance with beautifully ribbed bodies. Plum tomatoes are used traditionally in Italy for preserving and have an elongated, plum shape. Mini-plum tomatoes of the 'Santa' type have become very popular in recent years for snacking.

(In this book, mini-plums can be found in the Plum section, though other sources may classify them as cherries.) There are other shapes of tomato that do not fit easily into a category, in this book you'll find these in the Weird and wonderful section.

Colour

Although most people think of tomatoes as red, they can also be yellow, orange, green, white, nearly black or brown, and various shades of maroon-purple. The fruits may be a single colour, speckled or striped with a different colour, or multicoloured, for example 'Big Rainbow' (p63).

SEED CHAMBERS – also known as locules, these are usually constant in varieties

Bilocular fruits Most cherry tomatoes, such as 'Nectar' (p42), contain just two seed chambers.

Trilocular fruits Three seed chambers are common in standard varieties, such as 'Alaskan Fancy' (p66).

Multilocular fruits These are particularly common in beefsteak varieties, such as 'Ananas Noir' (p55).

The flavour

For many people, growing their own tomatoes is all about the flavour. Bought tomatoes often taste disappointing compared to home-grown crops, but the reasons behind this are complex.

Perceived flavour is derived from a combination of taste and smell. Over 400 volatile compounds have been identified in tomatoes, of which about 30 are thought to contribute to aroma.

The traditional sweet-sour taste of a tomato results from the sugar and organic-acid content of the fruit. Some tomatoes have a higher sugar content than others; for example, 'Black Cherry' contains twice as much sugar as the oxheart 'Sterling Old Norway'. Sugar content however can vary with season and the ripeness of the fruit. The flavour of tomatoes ripening in the increased sunlight of high summer is usually better than that of earlier crops.

For many years, commercial tomatoes have often been harvested when green and

Green Zebra (p26)

then exposed to ethylene gas to ripen them in storage. Tomatoes ripened on the vine are thought to have a much better flavour, although much of the aroma is released by the vine itself. Fruits nibbled while you work among the plants often taste best of all.

The tastiest of them all?

There is no doubt that flavour varies greatly between tomatoes. White and yellow fruits are generally less acid than red tomatoes. Many black- and brown-fruited plants are praised for their more complex flavours.

Varieties that are regularly commended as particularly flavoursome include the mini-plum 'Floridity' (p70), yellow cherry 'Snowberry' (p40), beefsteak 'Brandywine' (p59) in its various incarnations, the French 'Carmello', 'Green Zebra', and 'Black Prince'.

Black Cherry (p38)

The benefits

Tomatoes can play an important role in the diet. They are a valuable source of vitamins A and C, as well as several minerals, including calcium, iron, manganese, and, particularly, potassium.

Tomatoes contain an average of 0.09mg of vitamin A and 15.00mg of vitamin C per 100g (4oz) of fruit, as well as 397mg of potassium per 100mg of fruit. They also contain lycopene, which is a carotenoid (a pigment involved in photosynthesis) that gives red colouring to tomatoes, pink

Favorita
(p37)

grapefruit, and watermelons. Several population studies have indicated that diets high in lycopene may offer protection against certain cancers.

Lycopene in tomatoes can be absorbed more effectively by the body when the tomatoes have been processed in some way, particularly when they are combined with fat and heated – so drizzling olive oil over your tomatoes and roasting them should be especially beneficial.

A balanced diet

Lycopene is found in higher concentrations in red tomatoes; in studies, one cherry tomato of the variety 'Favorita' contained 1.39mg of lycopene, compared with 0.14mg found in a 'Golden Cherry' (p39) fruit. However, orange tomatoes have their own benefits – they have been found to contain much more vitamin A, in the form of beta-carotene, than red tomatoes.

There is no doubt that selecting and eating a variety of different tomatoes, of different shapes, sizes, and colours, as part of your diet will give you the best possible balance of nutrients. What more excuse do you need?

Aviro
(p66)

Best for sauces and salsas

Sauces and salsas are only as good as the tomatoes that go into them, so if you're planning to use your harvest to make up pots and pots of saucy delights, choose which tomato varieties you want to grow wisely. Here are some of the best of the bunch.

Sauces

Ripe plum tomatoes are the best for sauces because they have a good balance of flesh and juice. They also have a savoury quality that makes them very good for making flavoursome sauces. The plum 'San Marzano Lungo' is the traditional tomato of choice in Italian kitchens for superb sauces.

San Marzano Lungo (p65)

Floridity (p70)

Long Tom (p67)

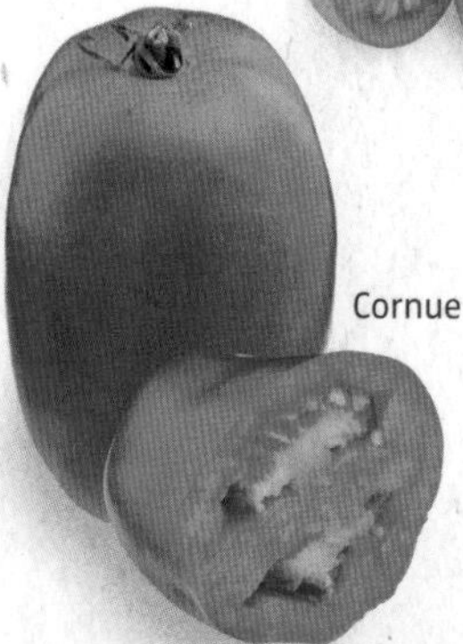

Juliet (p69)

Cornue des Andes (p72)

Salsas

For salsas, the choice of tomato depends on the salsa in question. If you are making fresh salsa, use a plump, meaty variety – beefsteaks are good. If you are making a chargrilled or cooked salsa, you can use other kinds of tomatoes, such as plums and standard globes.

Big Boy (p58)

Maskotka (p46)

Druzba (p32)

Stupice (p27)

Eva's Purple Ball (p31)

Mule Team (p25)

Best for soups

Because tomatoes for soups are likely to be puréed, you need to look for ripeness and depth of flavour – ideal requests when you are dealing with a glut. Most plum tomatoes are an ideal foundation for a hearty tomato soup, particularly the varieties shown here. Add some cherry tomatoes to the mix for extra sweetness and playfulness.

San Marzano Lungo (p65)

Ruby (p48)

Cornue des Andes (p72)

Rudolph (p70)

Principe Borghese (p66)

Rosada (p70)

Long Tom (p67)

Floridity (p70)

Tomatoberry (p46)

Loveheart (p41)

Nectar (p42)

Best for salads

For salads, both large plump beefsteaks and little pretty cherry tomatoes are ideal. Beefsteaks are fleshy and they soak up the flavour of a good vinaigrette. They also combine well with other ingredients, giving texture to a salad. Cherries of different colours look pretty in salads, add sweetness, and hold their shape.

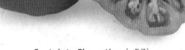

Costoluto Fiorentino (p57)

Supermarmande (p59)

Yellow Pigmy (p48)

Blue Ridge Mountain (p58)

Loveheart (p41)

Brandywine (p59)

Reisetomate (p77)

Riesentraube (p49)

German Pink (p61)

Golden Gem (p36)

Standard globe

When people think of tomatoes, it is usually the classic round tomatoes that come to mind. These generally vary between 70 and 100g (3 and 4oz) and have two to four seed cavities inside. They are more or less globe shaped with a regular outline and come in a rainbow of colours. The skin thickness, flesh texture, and flavour can also vary dramatically between varieties. Even fruits from a single plant can vary so don't be surprised to find unusual tomatoes among your crop.

Jubilee

⚘ ☀ **Heirloom, 72 days** Compact plants produce a good crop of orange tomatoes similar to the variety 'Sunray'. 'Jubilee' was an All-America Selections winner in 1943.

Characteristics Attractive, golden-yellow fruits with a somewhat dry flesh and low acidity. May grow to the size of a small beefsteak.

How to enjoy Often used for bottling and to make orange and tomato marmalade.

Roncardo

⚘ ☀ **F1 hybrid, 75 days** This plant from Holland produces large tomatoes in clusters of 4–6. Shows good disease resistance.

Large fruits not always as regular as this

Characteristics Large, red sweet fruits usually in the range of 100–150g (4–5oz) but some fruits may be larger beefsteak types.

How to enjoy Well-flavoured fruits are perfect sliced in sandwiches or burgers.

Ferline

⚘ ☀ ☀ **F1 hybrid, 75 days** Consistently heavy-cropping plant from France with excellent disease resistance, even outdoors.

Characteristics Solid, deep red tomatoes. They keep their texture well after picking. Sometimes classed as a beefsteak.

How to enjoy Good in sandwiches, as they remain firm after slicing.

Glacier

▽ ☀ ☀ **Heirloom, 65 days** Early potato-leafed plant introduced in Sweden in 1985. It is one of the most cold tolerant of all tomato plants. Produces a heavy crop.

Characteristics Flavourful, red globe tomatoes weighing just 50g (2oz) each. They have thick skins and very few seeds.

How to enjoy Better flavour than most early tomatoes, improves further grilled or roasted.

Few seeds

Shirley

⚘ ☀ 🏆 **F1 hybrid, 65 days** Very popular in England since the 1970s, 'Shirley' was first released in Israel. It is a reliable, early maturing plant, which produces more than 6kg (13lb) fruits. Ideal for unheated greenhouses or polytunnels.

Characteristics Very uniform, medium-sized fruits with a good texture and acidic taste.

How to enjoy Great for general kitchen use.

Very even-shaped fruits

Sungella

🌱 ☀ Heirloom, 70 days Raised in Norfolk, England, by a customer of Thompson and Morgan, who crossed the popular 'Sungold' (p41) with a larger fruited orange heirloom.

Characteristics Prolific crops of orange, golf ball-sized tomatoes, which are sweet and juicy.

How to enjoy Useful size to eat fresh but also good roasted with peppers and aubergines.

Bright orange, juicy fruits

Moneymaker

🌱 ☀ ☀ ☀ Heirloom, 70 days Reliable English outdoor tomato raised by F. Stoner in Southampton, also known for 'Stoner's Exhibition'. Strong plants are easy to grow in pots or the open ground.

Characteristics Good-looking, smooth, medium-sized, scarlet fruits.

How to enjoy Good for sauces and salsas.

Alicante

🌱 ☀ ☀ 🏆 Heirloom, 68 days English bred, reliable, early cropper that was introduced by Suttons Seeds in 1966 for growing outdoors or under cover. Easy to grow even in dull summers.

Characteristics These attractive, red tomatoes are resistant to greenback (pp118–19).

How to enjoy The firm-textured fruits are excellent baked or roasted.

Sioux

☀ ☼ ☀ ☀ **Heirloom, 60 days** Compact plants bred for use in the American plains by the Department of Horticulture, University of Nebraska in 1944, 'Sioux' also does well in the north of Britain and Europe.

Characteristics Small- to medium-sized, deep red tomatoes with a good flavour.

How to enjoy Try these well-flavoured tomatoes in salads early in the summer.

Tigerella

☀ ☀ ☼ ♉ **Heirloom, 60 days** Bred by Dr L. A. Derby in England around 1970. This early-maturing variety is sometimes called 'Mr Stripey', which is actually the name of an American striped beefsteak.

Characteristics Red-and-yellow-striped fruits with a tangy taste.

How to enjoy Attractive, tangy tomatoes for salads and sandwiches.

Red skins with light orange-yellow stripes

Mule Team

☀ ☼ **Heirloom, 80 days** A reliable, old plant from the USA that is high yielding and tolerant of many diseases.

Characteristics The bright red globes often show a slight ribbing around the stalk. They have a sweet and tangy taste.

How to enjoy Ideal for grilling and frying, and making delicious salsas.

Bright red fruits

Green Zebra

☀ ☼ ☼ Heirloom, **75 days** Introduced in 1983 by Thomas P. Wagner of Tater Mater Seeds in the USA.

Characteristics Yellowish green with deep green stripes, the attractive tomatoes have a tangy flavour and are resistant to splitting.

How to enjoy Serve cut into wedges as an interesting contrast to red tomatoes.

Fruits may vary in shape, usually less ribbed

Black Zebra

☀ ☼ Heirloom, **85 days** A child of Tom Wagner's 'Green Zebra' (above), stabilized by Jeff Dawson in California and first introduced in 2000.

Characteristics An attractive novelty tomato with deep red-and-green-striped fruits, which, sadly, are not as tasty as they look.

How to enjoy Aside from its decorative qualities, try 'Black Zebra' in soups and sauces.

Green-striped skin

Deep red flesh

Orange Pixie

○ ☼ Hybrid, **52 days** An attractive bushy plant growing to around 50cm (20in) in height only. It is an orange-fruited version of the red 'Pixie'.

Characteristics These distinctive orange fruits are slightly elongated globes with firm flesh and an excellent flavour.

How to enjoy Halve fruits and bake in the oven with a drizzle of olive oil.

Juicy fruits

The Amateur

♈ ☼ ☀ **Heirloom, 60 days** One of the best known British tomatoes, whose early cropping and good disease resistance makes it an excellent choice for beginners. Produces fruits reliably over a long season.

Characteristics Medium-sized, red tomatoes.

How to enjoy Useful everyday tomato for salads or frying.

Tamina

♈ ☼ **Hybrid, 60 days** A potato-leafed plant raised by the German company Saatzucht Quedlingburg. Plants produce very few sideshoots.

Characteristics These very even globe fruits are a bright red colour and have a good flavour.

How to enjoy Roast with plenty of olive oil, then sprinkle with feta cheese.

Stupice

♈ ☼ ☀ **Heirloom, 65 days** Potato-leafed plant that is very popular in its native Czech Republic. Seed was sent to the USA in 1977 by Milan Sodomka. Compact plants cope well in cold conditions.

Characteristics Usually round tomatoes with a rich flavour.

How to enjoy Well-flavoured tomatoes, excellent in soups, salsas, or stews.

Irregular-shaped fruits are fairly common

Essex Wonder

☀ ☀ ☀ **Heirloom, 75 days** Vigorous, potato-leafed English plant released by Dobie and Company for unheated greenhouse or outdoors. Ideal for planting in pots and growing bags, but can be rampant when planted in the ground.

Characteristics Firm-textured, scarlet fruits with a good flavour.

How to enjoy Ideal for grilling and frying.

Scotland Yellow

☀ ☀ ☀ **Heirloom, 70 days** Vigorous plants that were bred in Scotland for growing in cooler climates. Very similar to 'Yellow Ailsa Craig'.

Characteristics These round, golden-yellow, juicy tomatoes have a sweet and tangy taste, and stay in good condition long after picking.

How to enjoy These tasty tomatoes make excellent sauces.

Bright yellow-coloured fruits

Silvery Fir Tree

☀ ☀ **Heirloom, 58 days** A traditional Russian plant with distinctive, finely dissected foliage, making an ornamental patio plant.

Characteristics Very juicy, round, red tomatoes with a rather tart flavour and lots of seeds.

How to enjoy Serve grilled or fried as a breakfast tomato.

Celebrity

☷ ☼ ☼ **Hybrid, 70 days** Vigorous, high-yielding bushy plants showing good disease resistance. This variety was an All America Selections winner in 1984.

Characteristics Large, red tomatoes of reasonable flavour.

How to enjoy Roast with aubergines, peppers, and onions.

Pink Ping Pong

⅄ ☼ **Heirloom, 75 days** 'Ping Pong' is a Canadian variety bred in 1978 by the University of Manitoba in Winnipeg from 'Summit' and 'Red Cherry' (p40). The pink-fruiting strain is very prolific.

Characteristics The variety is so-named because the pink fruits are about the size of a ping-pong ball. Very sweet and juicy.

How to enjoy Ideal for snacking and salads.

Pink tinge to skin

Grosse Lisse

⅄ ☼ ☼ ☼ **Heirloom, 80 days** One of the most popular varieties in Australia, 'Grosse Lisse' is thought to have originated in Germany around 1870. The plant copes well with varying temperatures.

Characteristics The large, round fruits have thin skins and a meaty flesh with a pleasant, acidic taste.

How to enjoy Excellent in sandwiches.

Carters Fruit

⚔ ☀ **Heirloom, 75 days** Introduced by the seed company Carters of London, England in the 1930s. This is a strong-growing plant for indoors or outside.

Characteristics The skin of these crimson tomatoes has a distinctive powdery bloom.

How to enjoy These sweet, luscious fruits can be eaten as a dessert fruit.

Elegance

⚔ ☀ **F1 hybrid, 72 days** Widely grown commercially as on-the-vine tomatoes, this plant from the Dutch firm De Ruiter Seeds is sometimes offered to amateur growers as grafted transplants to increase vigour.

Characteristics Perfectly shaped, red tomatoes that are much favoured in taste trials.

How to enjoy Eat fresh in salads or use for richly coloured gazpacho.

Perfectly round even fruits

First In The Field

∀ ☀ ☀ **Heirloom, 60 days** A very productive tomato released in the early twentieth century that crops early and so often escapes attacks of blight. Plants are particularly cold tolerant.

Characteristics Traditional red globe tomatoes with a firm texture and good flavour.

How to enjoy Savour these well-flavoured tomatoes fresh early in the tomato season.

White Tomesol

🌱 ☀ Heirloom, **82 days** Of German origin, this is of a more regular shape than the similar 'White Queen'.

Characteristics Large, smooth-skinned, creamy white fruits of around 200g (7oz).

How to enjoy Some people find white tomatoes unappealing but this variety has a pleasant citrus-like taste and can be eaten like an apple.

Fairly dry flesh

Golden Sunrise

🌱 ☀ 🏆 Heirloom, **72 days** Introduced by the English seed company Carters in the early 1890s, 'Golden Sunrise' is a popular classic tomato.

Characteristics The smooth, golden-yellow tomatoes have a low acidity, thin skins, and firm flesh.

How to enjoy A sweet-tasting, multi-purpose salad tomato.

Eva's Purple Ball

🌱 ☀ ☀ Heirloom, **70 days** An old German heirloom that was taken to New Jersey, USA, in the late nineteenth century by the family of Joseph J. Bratka. High-yielding plants show some tolerance of blight.

Characteristics Extremely sweet and juicy fruits are pinkish red (not purple) and evenly round when ripe.

How to enjoy Grill for breakfast or make salsas.

Druzba

☀ ☼ Heirloom, 80 days The word *druzhba* means friendship in Bulgarian. These disease-resistant plants reliably produce a prolific crop.

Characteristics These deep red globes have a well-balanced flavour.

How to enjoy Juicy fruits ideal for making into rich savoury sauces and salsas.

Earl of Edgecombe

☀ ☼ ☼ Heirloom, 73 days The seventh Earl of Edgcumbe was a New Zealand sheep farmer before claiming his title in 1965. He took this tomato to England with him.

Characteristics Uniform, deep orange globe tomatoes with a good flavour and not too many seeds.

How to enjoy Delicious raw but the firm, meaty flesh is also good for grilling and frying.

Rich orange-coloured skin and flesh

Ailsa Craig

☀ ☼ Heirloom, 80 days First introduced in 1912 by Alexander and Brown of The Scottish Seed House, 'Ailsa Craig' was raised by Alan Balch as a cross between 'Fillbasket' and 'Sunrise'. Named after an isle in the Firth of Clyde.

Characteristics Blood red, smooth tomatoes, which are firm and juicy but fairly acidic.

How to enjoy Versatile; good for any dish.

Totem

☐ ☀ Hybrid, 75 days Bred by the British company Floranova in 1996, this is a neat, very productive stocky bush ideal for pots or window boxes.

Characteristics Bright red tomatoes that can be rather small for standard globes. Sometimes classed as a cherry tomato.

How to enjoy Grow on a windowsill to have a ready supply of delicious snacking tomatoes.

Czech's Excellent Yellow

🌱 ☀ Heirloom, 72 days Introduced by Ben Quisenberry of Ohio in 1976, seed of this variety probably came from Milan Sodomka of the Czech Republic, who sent seed of other varieties including 'Czech's Bush'.

Characteristics These very smooth, round fruits are an attractive rich yellow colour.

How to enjoy The soft-textured tomatoes are lovely when used in chutneys and pickles.

Long Keeper

⩌ ☀ Heirloom, 78 days A semi-bush plant released by Burpee Seeds in the USA. The tomatoes ripen slowly for 1–3 months after harvest.

Characteristics Orange-red tomatoes will store for 12 weeks or more in a cool, dry place.

How to enjoy Not the best flavoured tomato but convenient for winter use. Best cooked in soups and casseroles.

Distinctive orange-red skin

Cristal

☝ ☀ 🏆 **F1 hybrid, 85 days** High-yielding plants that are very resistant to disease.

Characteristics Glossy, bright red tomatoes up to 120g (4½oz) in weight with a richly coloured flesh and good flavour.

How to enjoy Excellent variety for eating in fresh salads and sandwiches.

Siberian

⊻ ☀ ☀ **Heirloom, 48 days** Russian heirloom variety that sets fruit even at low temperatures and is productive even in short seasons.

Characteristics Bright red globe tomatoes.

How to enjoy Use in soups and stews or the first summer salads.

Rich red, succulent fruit

Sub-Arctic Plenty

◯ ☀ ☀ **Heirloom, 50 days** Bred at the Beaverlodge Research Station in Alberta, Canada, in the 1970s, specifically for short season gardening in cool conditions. Sister varieties include 'Sub-Arctic Delight', 'Sub-Arctic Maxi', and 'Sub-Arctic Midi'.

Characteristics Bright scarlet fruits with a sweet and tangy taste.

How to enjoy Lovely roasted with thyme.

Lemon Boy

☀ ☼ **F1 hybrid, 72 days** Introduced by Petoseed of the USA in 1984, the disease-resistant plants are generally very prolific.

Characteristics Large, bright lemon-yellow tomatoes with a low-acid flavour and not too many seeds.

How to enjoy Try using for colourful salsas and gazpacho.

Few seeds

Harzfeuer

☀ ☼ **F1 hybrid, 65 days** This early cropping plant from Germany is very popular with market gardeners. The name translates as "resin fire". Open-pollinated seed is sometimes listed.

Characteristics Large, round, red fruits have a slightly acidic flavour.

How to enjoy Good general purpose tomato for salads or cooking.

Flamme

☀ ☼ **Heirloom, 68 days** Also known as 'Jaune Flamme', this is a prolific-cropping old and much-loved French plant.

Characteristics Beautiful, smooth, round fruits of a rich tangerine orange.

How to enjoy Wonderful grilled or roasted but also very good for drying.

Cherry

Small, roundish tomatoes are referred to as cherry tomatoes, as they resemble in shape and size the fruit of that name. They generally weigh in the region of 10–20g (½–1oz), are usually roughly spherical with just two seed chambers, and have a high sugar content. Most plants are easy to grow and trouble-free. The tiny fruits related to the wild species *Solanum pimpinellifolium* are called currant tomatoes. They weigh around 2g (¹⁄₁₆oz) and have a particularly sweet taste.

Tumbler

▽ ☼ **Hybrid, 55 days** German-bred trailing plant, which crops over a long period. One of the best tomato plants for cascading out of a hanging basket or a window box.

Characteristics Bright red cherry tomatoes with a sweet taste.

How to enjoy Ideal cooked on skewers for dishes such as shish kebabs.

Golden Gem

⌇ ☼ **F1 hybrid, 65 days** A prolific tomato bred in China, which may have 20 to 70 fruits in a cluster.

Characteristics The golden-yellow fruits have a good flavour and a high sugar content of around 10 per cent. They show some resistance to splitting.

How to enjoy An ornamental cherry tomato – try using it to decorate a fruit salad.

Golden-yellow fruits

Tiny Tim

○ ☀ **Hybrid, 50 days** Fast-maturing plants introduced in 1944, ideal for growing in window boxes or pots. Plants will fruit successfully in 15cm (6in) pots.

Characteristics The small, sweet tomatoes are reliably produced very early in the season.

How to enjoy Perfect patio plants, you can enjoy the fruits while relaxing in the sun.

Gartenperle

🌿 ☀ **Heirloom, 68 days** 'Gartenperle', translated as "garden pearl", is of German origin. It forms a dwarf-trailing bush ideal for growing in hanging baskets or patio pots and produces fruits in great profusion.

Characteristics These are small, sweet, pinkish red fruits.

How to enjoy Very popular variety with moreish fruits to nibble off the plant.

pinkish fruits

Favorita

🌿 ☀ 🏆 **F1 hybrid, 65 days** Vigorous disease-resistant plants, which will usually produce a good crop even in wet summers.

Characteristics Rich, red cherry tomatoes held on long trusses. They have tough skins and a fairly acidic taste, which becomes sweeter in hot summers.

How to enjoy Cook in flans and quiches.

Fox Cherry

☀ ☼ **Heirloom, 80 days** This popular tomato is produced in large quantities over a long season on tall, disease-resistant plants.

Characteristics Large, red cherries weigh around 30g (1¼oz) each and have a rich, slightly acidic flavour.

How to enjoy Serve on skewers with mini mozzarella balls and an aïoli dressing.

Relatively large cherries

Tommy Toe

☀ ☼ ☀ ☼ **Heirloom, 70 days** From the Ozark Mountains in Arkansas, USA. Widely distributed in Australia by Diggers Seeds, these long-cropping plants show good blight resistance.

Characteristics Voted top in taste trials, the large cherries have an excellent flavour.

How to enjoy Eat freshly picked from the vine or in salads.

Black Cherry

☀ ☼ ☀ **Heirloom, 65 days** Vigorous, high-yielding plants developed by Vince Sapp in the USA. Released in 2003.

Characteristics Produces large clusters of dusky purplish, round tomatoes with a rich flavour and high sugar content. Can be prone to splitting.

How to enjoy Eat straight off the vine or mix with other cherry varieties for a colourful salad.

Chocolate Cherry

🌱 ☀️ **Heirloom, 70 days** Raised in the USA, these plants produce masses of fruits, usually in clusters of eight.

Characteristics Looks similar to 'Black Cherry' (opposite) but is marginally larger and more resistant to splitting. Good flavour and a deep maroon-black colour.

How to enjoy Lower in calories than real chocolate, this is the ideal sweet snack.

Thick skins

Golden Cherry

🌱 ☀️ **F1 hybrid, 60 days** Introduced by Suntech Seeds of Taiwan, this is a vigorous plant for greenhouses or outdoors. Produces long trusses of well-spaced fruits.

Characteristics These firm cherries are bright orange-yellow and have a very sweet, tangy taste.

How to enjoy The fruit's excellent flavour makes it particularly appealing to children.

Matt's Wild Cherry

🌱 ☀️ **Heirloom, 70 days** An early cropping currant of Mexican origin that was named after Matt Liebman. It is a vigorous sprawling plant, which will self-seed in the garden.

Characteristics The tiny, flavourful, red tomatoes weigh less than 3g (¹⁄₁₆oz) each. They have a firm texture and contain many seeds.

How to enjoy These exceedingly sweet and juicy fruits are best scattered over salads.

Plenty of seeds

Red Cherry

🌱 ☀ **Heirloom, 65 days** An RHS report in 1877 noted that this tomato was a profusely fruiting, ornamental variety.

Characteristics Pinkish red, round fruits with a good sweet flavour. Some suppliers use this name for a mini-plum variety.

How to enjoy Popular as a snack or oven roasted with peppers and aubergines.

Glossy red skins

Snowberry

🌱 ☀ **F1 hybrid, 75 days** Raised in The Netherlands by Sahin Seeds, this plant produces a heavy yield of fruits.

Characteristics The relatively large, attractive fruits are creamy yellow when mature and have a rich, sweet flavour with citrus hints.

How to enjoy Ideal for nibbling on as a snack between meals.

Creamy-yellow fruits

Jelly Bean

🌱 ☀ **Hybrid, 66 days** Introduced in the USA in 2007, there are red and yellow versions of this variety.

Characteristics Sweet, grape-like glossy fruits weighing about 20g (1oz) with 15–30 in a cluster. Good resistance to splitting and very long keeping.

How to enjoy Straight off the plant for a garden snack or will dry well for winter use.

Bright glossy fruits

Loveheart

🌱 ☀️ **F1 hybrid, 75 days** Also sold as Cutie, this variety from Taiwan was first released in 2005. The fruits increasingly take on a heart shape as the plant matures.

Characteristics The rich red, heart-shaped fruits make this a very appealing choice, especially as they have a good tangy flavour.

How to enjoy Serve at a special dinner for the one you love!

Distinctly heart-shaped fruits

Sunset

🌱 ☀️ 🏆 **F1 hybrid, 65 days** Productive hybrid that has been successful in RHS trials. Healthy vines produce a good yield.

Characteristics Bright orange fruits are sweet and richly flavoured.

How to enjoy The very ornamental fruits look particularly dramatic in a salad with 'Black Cherry' (p38) or 'Brown Berry' (p45).

Sungold

🌱 ☀️ ☀️ 🏆 **F1 hybrid, 60 days** Bred in Japan by the Tokita Seed Company, 'Sungold' was released in Britain and America in 1992 and rapidly became the most popular cherry tomato. The plants are very productive and have good virus resistance.

Characteristics Very sweet, golden-yellow, thin-skinned tomatoes.

How to enjoy Irresistible straight off the vine.

Thin skins

Sakura

F1 hybrid, 70 days A modern high-yielding and early fruiting plant raised in The Netherlands. Plants produce long trusses of approximately 20 tomatoes.

Bright red fruits

Characteristics These uniform bright red cherries have a very sweet flavour.

How to enjoy Firm tasty fruits hold well on skewers to make savoury kebabs or to dip in cheese fondue.

Nectar

F1 hybrid, 68 days Modern early fruiting hybrid, from the seed company Enza Zaden in Holland, that grows best under glass.

Glossy skins

Characteristics Tasty, glossy fruits on long trusses store well after harvest.

How to enjoy Good for roasting – bake in the oven with sausages, herbs, and garlic. Alternatively, add to a soup.

Gardener's Delight

Heirloom, 75 days One of the best known of all tomatoes, 'Gardener's Delight', or 'Sugar Lump', is of German origin, bred by Paul Tellhelm and introduced in 1950. The plant is a reliable producer.

Characteristics Perfect red cherry tomatoes weighing around 25g (1oz), with an excellent traditional tomato flavour.

How to enjoy Eat handfuls as a healthy snack.

Striadel

🌱 ☀ **Hybrid, 75 days** One of the Del series raised by Lewis Derby of the Glasshouse Crops Research Institute in England. Bred from 'Gardener's Delight' (opposite), others in the series include the yellow 'Daffodel', pink 'Rosadel', and pale yellowish 'Albadel'.

Characteristics Red cherries with the same sweet tangy flavour as their parent.

How to enjoy Eat in a colourful salad.

Skins faintly striped gold

Velvet Red

🌱 ☀ **Heirloom, 78 days** Thought to be synonymous with 'Angora Super Sweet', this is one of a few plants that has beautiful silvery foliage, which deters pests.

Characteristics Noted for the silver hairs on the leaves and the slightly fuzzy fruits, which taste very sweet.

How to enjoy Eat straight off the vine while you are stroking the leaves.

Fine hairs on stalk

Sparsely hairy fruits

Micro-Tom

◯ ☀ **Heirloom, 65 days** Developed by Scott and Harbaugh at the University of Florida who have also released the yellow-fruited 'Micro-Gold'. Particularly tiny plants, ideal for planting in a pot on a windowsill.

Characteristics Miniature, red fruits with a firm texture and sweet taste.

How to enjoy Best eaten freshly picked from the plant.

Jasper

▽ ☼ ◐ ♈ **F1 hybrid, 68 days** These
semi-bush plants produce high yields of
firm fruits and show good disease
resistance and heat tolerance.

Characteristics Traditional red cherries
in appearance, these fruits have quite a
distinctive acidic taste.

How to enjoy Ideal with mozzarella in salads.

Balconi Red

◯ ☼ **Hybrid, 70 days** Small bushy
plants bred by Saatzucht Quedlinburg
of Germany. They produce trailing stems
and are therefore ideal for growing in
baskets or window boxes.

Characteristics Small, sweet cherry tomatoes.

How to enjoy Grow with the yellow variety
(below) and enjoy the contrasting colours.

Balconi Yellow

◯ ☼ **Hybrid, 70 days** The yellow-fruited
version of this popular German variety is
an ideal companion for creating colourful
containers on a balcony or patio.

Characteristics Very sweet, bright yellow
cherry tomatoes.

How to enjoy Use with the red variety (above)
for colourful salads.

Decorative, brightly coloured fruits

Suncherry Premium

🌱 ☀ 🏆 **F1 hybrid, 75 days** An early ripening prolific plant bred in Japan. 'Suncherry Extra Sweet' is part of the same series.

Characteristics These uniform, red tomatoes have very glossy skins and a particularly good sweet flavour.

How to enjoy Use with 'Sungold' (p41) in colourful quiches and flans.

Brown Berry

🌱 ☀ **Heirloom, 75 days** Bred by Sahin Seeds of The Netherlands, this plant produces a high yield of attractive tomatoes with an unusual colour.

Characteristics These distinctive brown tomatoes are not as sweet as those of 'Black Cherry' (p38) or 'Snowberry' (p40).

How to enjoy Best served as a rainbow salad with other colours of cherry tomato.

Very juicy fruits

Pepe

🌱 ☀ 🏆 **F1 hybrid, 68 days** Very productive plants from the European division of the Japanese Takii Seed Company, with 35–50 fruits in a cluster. High disease resistance.

Characteristics Mid-sized cherry tomatoes with a very sweet taste.

How to enjoy Perfect for nibbling straight off the vine.

Extra Sweetie

❦ ☀ ♛ **F1 hybrid, 70 days** Excellent variety similar to 'Sweet 100' and 'Sweetie', it produces grape-like clusters of fruits.

Characteristics Bright red tomatoes of an intermediate mini-cherry/grape shape. They have a particularly high sugar content.

How to enjoy Use in children's packed lunches as an easy way to encourage them to eat fruit.

Maskotka

◯ ◐ ☀ **Hybrid, 70 days** Compact plants from Poland, the name means "mascot". Ideal for growing in containers, as the tomatoes will tumble over the edge.

Characteristics These sweet fruits weigh 25–35g (1–1½oz) and show good split resistance.

How to enjoy Why not use to make a tasty sauce to accompany Polish golabki – stuffed cabbage rolls?

Particularly juicy fruits

Tomatoberry

❦ ☀ **F1 hybrid, 60 days** This variety was introduced in 2007 by the Japanese Tokita Seed Company who also released the very popular 'Sungold' (p41).

Characteristics The bright red fruits are shaped like strawberries, giving them a distinctive look and just like strawberries, they are particularly sweet and juicy.

How to enjoy Eat as a snack or use in a soup.

Strawberry-shaped fruits

Green Tiger

🌱 ☀ **Hybrid, 72 days** A recent open-pollinated variety, sold throughout supermarkets in the UK.

Characteristics Attractive deep burgundy-coloured fruits with olive-green stripes. The glossy skins are thick and resistant to splitting.

How to enjoy Best roasted, as they are too tough for eating raw.

Deep green stripes on skin

Vivid green seeds

Golden Pearl

🌱 ☀ **Heirloom, 60 days** This plant produces grape-like clusters of miniature yellow fruits over a long season.

Characteristics The golden-yellow, pea-sized tomatoes have little flesh but large seeds for the size of the fruits.

How to enjoy Best served scattered over a savoury dish for their ornamental value.

Picolino

🌱 ☀ **F1 hybrid, 70 days** A short-growing but vigorous plant from Holland. Commercially, at least two flowers per truss are removed so the remaining fruits develop as uniformly as possible.

Characteristics Cocktail-type tomato with firm, glossy red fruits averaging 30g (1¼oz) each.

How to enjoy They can seem a little too firm to eat fresh but are lovely roasted.

Red Pigmy

�’ ☀ **Heirloom, 65 days** This early fruiting plant was introduced by K. Sahin in Holland. Plants grow up to 35cm (14in) tall with a trailing habit ideal for hanging baskets. Unusual leaves have a bumpy surface.

Characteristics Bright red cherry tomatoes sometimes have a yellow tint near the stalk.

How to enjoy Serve in salads with the yellow variety (below) for contrast.

Sometimes have a yellow tint around stalk

Yellow Pigmy

�’ ☀ **Hybrid, 65 days** Released by Sahin Seeds in Holland as a sister variety to the popular 'Red Pigmy' (above).

Characteristics Bright yellow cherry tomatoes. Like other yellow sister varieties, they are slightly sweeter then their red counterpart.

How to enjoy The small sweet fruits are best eaten whole in salads.

Ruby

☘ ☀ ☀ ♔ **F1 hybrid, 60 days** Vigorous plant from Taiwan produces many long trusses, which each have 25–30 tomatoes. More blight resistant than many varieties.

Characteristics Red fruits, which have a very rich flavour and a particularly sweet and succulent taste.

How to enjoy Perfect for adding colour and flavour to Greek salads or try in a soup.

Tumbling Tom Red

☰ ☼ **Hybrid, 78 days** Introduced in England in 2002 by Floranova, these high-yielding plants have a cascading growth habit, making them ideal for containers.

Characteristics Small, sweet, red tomatoes.

How to enjoy Use with 'Tumbling Tom Yellow' (below) and 'Green Grape' to make a traffic lights salad.

Tumbling Tom Yellow

☰ ☼ **Hybrid, 78 days** The sister to 'Tumbling Tom Red' (above) with the same cascading habit and productivity. Grow in hanging baskets or raised beds on a patio.

Characteristics Small, sweet, yellow tomatoes.

How to enjoy Steal off the bush as you tend to your plants.

Riesentraube

🌱 ☼ ☼ **Heirloom, 80 days** Roughly translated as "bunch of grapes", this German variety, grown widely in Pennsylvania in the nineteenth century, was made popular by Curtis D. Choplin of South Carolina in the 1990s.

Characteristics Huge bunches of red tomatoes with pointed ends. Excellent flavour.

How to enjoy Usually eaten fresh in salads but used traditionally to make tomato wine.

Nipple-shaped pointed ends

cherry **49**

Piccolo

🌱 ☀ 🏆 **F1 hybrid, 78 days** Registered in 2007 by the French seed company Gautier Semences, these sprawling plants are easily grown and very disease resistant.

Characteristics Extremely tasty, these tiny, bright red tomatoes are very popular for their intense sweet taste.

How to enjoy Eat them quickly from the vine, before anyone else beats you to them.

Cherrola

🌱 ☀ 🏆 **F1 hybrid, 72 days** Productive plant with long trusses of up to 20 well-spaced fruits. Highly praised in the RHS's trial of 2007, the first retail release in the UK was for the 2008/9 season.

Characteristics Produces ornamental clusters of bright red cherries with an excellent flavour.

How to enjoy Use whole in flans and quiches. They are great combined with goat's cheese.

Amoroso

○ ☀ **F1 hybrid, 72 days** A cocktail-type, popular commercial variety from Rijk Zwaan of The Netherlands. Compact plants, which show good disease resistance, produce evenly spaced trusses of fruits.

Characteristics Shiny red fruits have a high sugar content.

How to enjoy Roast whole with other Mediterranean-style vegetables.

Large cherries

Sweet Pea

🌱 ☀ **Heirloom, 78 days** This prolific currant tomato was produced in Holland. The plants have delicate foliage and produce hundreds of pretty red fruits for harvesting on the vine.

Characteristics Tiny, currant-sized tomatoes just 2g (¹⁄₁₆oz) each have a rich flavour.

How to enjoy Tasty, decorative fruits make jewel-like additions to salads.

Tiny, seedy fruits ⟶

Broad Ripple Yellow Currant

🌱 ☀ ☀ **Heirloom, 75 days** Saved from a plant that was found growing in a paving crack in the Broad Ripple district of Indianapolis, USA.

Characteristics Prolific producer of small, sweet, low-acid, yellow tomatoes weighing just 8–10g (¹⁄₂oz) each.

How to enjoy Perfect for children to eat off the vine like sweets.

Gold Rush

🌱 ☀ **Heirloom, 78 days** Like its sister, 'Sweet Pea' (above), this prolific currant tomato was also produced in Holland. Harvest as a whole truss.

Characteristics The tiny but particularly richly flavoured, sweet orange berries stay on the vine for prolonged periods without splitting.

How to enjoy Very decorative, tasty fruits, ideal for garnishing sweet or savoury dishes.

Very small fruits

Beefsteak

Beefsteak is probably the most commonly used word to describe large tomatoes that usually have five or more locules. The tomatoes typically weigh around 180–250g (6–9oz) but may be huge, weighing as much as 1kg (2.2lb). They usually have a slightly flattened look, being wider than they are long, and may seem misshapen. There are a number of varieties, however, that have beautifully fluted outlines. Their meaty flesh makes them ideal for slicing and eating in sandwiches.

Country Taste

☀ **F1 hybrid, 78 days** One of relatively few hybrid beefsteak varieties, 'Country Taste' is from the Dutch firm Nunhems Zaden. Shows good resistance to disease.

Characteristics These are meaty, deep red fruits of a good flavour.

How to enjoy Excellent slicing variety for enjoying with burgers and a convenient size for general culinary use.

Dr Wyche's Yellow

☀ **Heirloom, 85 days** Dentist Dr John Wyche of Cherokee heritage, was part owner of a circus. He used elephant manure to fertilize his Oklahoma tomato garden.

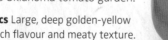

Meaty flesh with few seeds

Characteristics Large, deep golden-yellow fruits with a rich flavour and meaty texture.

How to enjoy A lovely variety to slice and serve in sandwiches or makes a well-flavoured yellow sauce to serve with fish.

Pineapple (Ananas)

☂ ☀ **Heirloom, 85 days** This potato-leafed plant is a popular heritage variety from Kentucky in the USA.

Characteristics The yellow fruits blush red as they ripen and can reach 1kg (2.2lb) in weight. They have a mild, fruity flavour.

How to enjoy The firm flesh is sweet and delicious served simply in wedges with cucumber and melon.

Sometimes has red blush

Aunt Ruby's German Green

☂ ☀ ☀ **Heirloom, 80 days** Ruby Arnold of Greenville, Tennessee, obtained the seed of this tomato from her German grandfather. Be sure to isolate the plant if you are saving seed.

Characteristics Spicy-sweet, large fruits, weighing 400g (14oz) remain green when ripe.

How to enjoy Best served fresh rather than cooked to appreciate the unusual taste.

Hillbilly Potato Leaf

☂ ☀ **Heirloom, 85 days** A late-season gourmet variety from Ohio. The potato-leafed plants are particularly vigorous.

Characteristics The sweet fruits ripen to a mottled-golden colour streaked with red, and can reach around 1kg (2.2lb).

How to enjoy The well-flavoured fruits are best served fresh in thick slices.

Attractive-patterned flesh

Cuostralée

🌱 ☀ **Heirloom, 85 days** Huge plants produce equally impressive fruits of this popular French variety. In wet summers, grow under cover, as they can be particularly susceptible to foliage diseases (p116).

Characteristics The red, intensely flavoured tomatoes can weigh up to 1–1.5kg (2–3lb).

How to enjoy Each tomato could be a meal in itself but are best shared with friends.

Many seed chambers

Faworyt

🌱 ☀ ☀ **Hybrid, 70 days** Meaning "favourite" in Polish, these compact-growing plants were raised by PlantiCo of Poland. They perform well in various growing conditions.

Characteristics These large, sweet fruits of around 400g (14oz) are a raspberry-pink colour.

How to enjoy Delicious sweet-fleshed variety to be savoured fresh.

Few seeds

Halladay's Mortgage Lifter

🌱 ☀ **Heirloom, 90 days** Marshall Cletis Byles sold his own strain of this variety, 'Radiator Charlie's Mortgage Lifter', to pay off his mortgage. This strain comes from relatives of James Halladay of Pennsylvania.

Characteristics These usually weigh 400–800g (14oz–1¾lb), and are resistant to splitting.

How to enjoy Savour in slices or sell to pay off your own mortgage.

Legend

▽ ☀ ☀ 🏆 **Hybrid, 68 days** Bred by Baggett and Myers at Oregon State University. Sets earlier fruit than the related 'Oregon Spring'. Excellent for cool climates and notable for its blight resistance.

Characteristics Variable-sized fruits that may exceed 400g (14oz), with few seeds.

How to enjoy Delicious grilled with a drizzle of olive oil.

Ananas Noir

🌿 ☀ **Heirloom, 85 days** Translated as "black pineapple", this high-yielding French plant introduced in 2005 by Pascal Moreau won't always come true from seed.

Characteristics These large, multicoloured fruits can weigh around 600g (20oz). They have a rich taste described as smoky with citrus hints.

How to enjoy To appreciate the rainbow of colours and distinct flavour, serve in salads.

Green, red, and yellow flesh

Pink Accordion

🌿 ☀ ☀ **Heirloom, 78 days** Very popular heritage variety, admired for the unusual shape of the tomatoes.

Hollow fruits

Characteristics The large, pinkish red fruits are ruffled like an accordion. They have a mild, sweet flavour.

How to enjoy Fruits are easy to hollow out and look beautiful stuffed.

Jack Hawkins

🌱 ☀️ **Hybrid, 72 days** A commercial variety bred in the USA, these plants are much grown in the UK for the supermarket trade.

Characteristics The large, well-flavoured fruits have a particularly sweet flesh.

How to enjoy Slice to serve on burgers or use to make a wonderful soup.

Pantano

🌱 ☀️ **Heirloom, 70 days** *Pantano* is Italian for "marsh" and this productive plant, also known as 'Pantano Romanesco', is said to have been grown on the former marshes near Rome.

Characteristics These large, scalloped fruits are usually green around the shoulders.

How to enjoy The thick, tasty flesh and few seeds make this an excellent slicing tomato.

Costoluto Genovese

🌱 ☀️ ☀️ **Heirloom, 78 days** *Costoluto* means "ribbed" in Italian and these big, beautifully ruffled tomatoes were one of the first kinds of tomatoes to be introduced to Europe in the sixteenth century.

Ribbed fruits

Characteristics These attractive, scalloped, cherry-red fruits have a firm, meaty texture.

How to enjoy Traditionally used in Italy for purées, they are ideal for juicing too.

Costoluto Fiorentino

⚑ ☼ ☀ 🏆 **Heirloom, 80 days** A variety from Tuscany in Italy, closely related to but usually not as ruffled as 'Costoluto Genovese' (opposite).

Characteristics These juicy, bright red fruits are variable in shape, usually slightly ribbed.

How to enjoy Wonderful on pizzas or use to make flavourful risottos and to add a meaty component to salads.

Attractive-patterned flesh

Amana Orange

⚑ ☀ 🏆 **Heirloom, 80 days** Introduced in 1985 by Gary Staley of Florida, who named it after the Amana Corporation where he worked as a customer service manager.

Characteristics Big and beautiful, the shiny orange, ribbed fruits have a mild taste.

How to enjoy Admire sliced thinly on a plate, sprinkled with torn basil leaves.

Solid flesh with few seeds

Delicious

⚑ ☀ **Heirloom, 78 days** Introduced by Burpee Seeds in the USA in 1964. This variety holds the Guinness World Record for the heaviest tomato ever grown.

Characteristics Fruits average 800g (1.7lb). They can be very sweet but sadly the biggest fruits do not necessarily live up to their name.

How to enjoy Use for soups and sauces or to make an attempt on that world record (p126).

Big Boy

☀ ☀ **F1 hybrid, 78 days** Bred by
Oved Shifriss and released in 1949 by
Burpee Seeds in the USA, 'Big Boy' is a
much-loved variety still in production as
an F1 hybrid. Others in the series, including
'Ultra Boy', 'Ultra Girl', and 'Early Girl', show
better disease resistance.

Characteristics Large, bright red fruits.

How to enjoy Use to make excellent ketchup.

Better Boy

☀ ☀ **F1 hybrid, 72 days** Improved
version of 'Big Boy' (above), the plant has
greater disease resistance and productivity.
The plant holds the world record for the
most tomatoes produced by a single plant,
at 155kg (340lb).

Characteristics Large, evenly shaped fruits.

How to enjoy Perfect for slicing in sandwiches.

Blue Ridge Mountain

☀ **Heirloom, 82 days** A potato-leafed
plant from the Blue Ridge Mountain region
of North Carolina in the USA.

Characteristics Well-flavoured sweet fruits
are pinkish red and weigh around 600g (20oz).

How to enjoy Ideal for making sauces and
salsas or use to add flavour to salads.

*Richly coloured
fruits with a
pink tinge*

Supermarmande

☿ ☼ **Heirloom, 62 days** 'Marmande', named after the area of France from where the variety came, is a very popular, old semi-bush plant. 'Supermarmande' crops earlier and shows more disease resistance. 'Rouge de Marmande' (Burkes Backyard) is popular in Australia.

Characteristics Large, flavoursome fruits.

How to enjoy Use in soups, salads, and coulis.

Lightly ribbed, cherry-red fruits

Brandywine

🌿 ☼ **Heirloom, 85 days** Introduced in 1889, this extremely popular tomato was named after the Brandywine River in Pennsylvania. Originally blood red, there are now many strains of different colours.

Characteristics These large fruits (shown here enlarged but not in proportion) in the 400–800g (14oz–1.7lb) range have a characteristic sweet taste.

How to enjoy Wonderful served fresh in slices.

Often cracks at shoulders

Dense flesh

Evergreen

⚘ ☀ **Heirloom, 75 days** Introduced by Glecklers Seedmen of Ohio in 1956, this variety is also called 'Emerald Evergreen' and 'Tasty Evergreen'.

Characteristics Easily peeled, greenish yellow fruits, which have a lovely citrus-like flavour.

How to enjoy Excellent when used half-and-half with apples in open tarts.

Purple Calabash

⚘ ☀ ☀ **Heirloom, 85 days** Originating in Texas, 'Purple Calabash' was first released commercially in 1987 by Glecklers Seedmen and others.

Characteristics A most unusual heavily ribbed tomato that is an intriguing rich purplish red colour like a well-ripened loganberry. Has a tangy and somewhat spicy flavour.

How to enjoy Best used in casseroles.

Deeply ruffled fruits

Kellogg's Breakfast

⚘ ☀ **Heirloom, 79 days** Named after Darrell Kellogg of Michigan, USA, not the purveyor of breakfast cereals, this is a popular and prolific plant.

Characteristics These large, vibrant orange fruits are somewhat irregular in shape and have thin skins.

How to enjoy Use to make rich-flavoured and brightly coloured juices.

Vibrant orange fruits

German Pink

☀ ☀ Heirloom, 85 days A potato-leafed plant taken to America in 1883 from Bremen, Germany, by Michael Ott.

Characteristics The large, pinkish red fruits weigh over 400g (14oz) and contain few seeds.

How to enjoy The fruits are traditionally eaten in slices, sprinkled with sugar. Alternatively, enjoy in a salad.

Meaty flesh with few seeds

Black Russian

☀ ☀ ☀ Heirloom, 78 days A mid-season variety from Russia. Popular in Australia, as it copes well with the climate.

Characteristics These deep maroon-coloured fruits often remain green at the shoulders. Good tender flesh but the fruits can be prone to splitting.

How to enjoy Include sliced in a salad with red and green varieties for colour contrast.

Often remains green at shoulders

Rose de Berne

☀ Heirloom, 75 days Attractive rosy-pink tomato from Switzerland. The name translates as "rose from Berne", Switzerland's capital.

Characteristics Variable in shape and size from 150 to 200g (5 to 7oz) but always with an intense sweet taste and a low acid content.

How to enjoy Serve in thick slices with cheese or cold meats.

Soft, meaty flesh

Thin skin

Gold Medal

🌱 ☀ **Heirloom, 85 days** Introduced in the USA as 'Ruby Gold' by John Lewis Childs in 1921, this variety has had several name changes and was also called 'Early Sunrise'.

Characteristics The large, sweet fruits (shown here enlarged but not in proportion) are orange-yellow streaked throughout with red.

How to enjoy Eat fresh or use to make a wonderful refreshing juice.

Orange flesh streaked with red

Sweet and particularly soft flesh

Aker's West Virginia

🌱 ☀ **Heirloom, 88 days** A family heirloom from Carl Aker of West Virginia, home to many other heritage varieties including the pink-fruited 'Tappy's Finest' and 'West Virginia Straw'. Plants are high yielding and disease resistant.

Characteristics Large fruits with a rich flavour.

How to enjoy Excellent for slicing.

May show concentric cracks around stalk

Big Rainbow

🌿 ☀ **Heirloom, 85 days** Named by Dorothy Beiswenger of Minnesota in 1983, this is a huge red and yellow tomato. It certainly lives up to its name. In cool climates grow under glass.

Characteristics These large, colourful, meaty fruits average around 600g (20oz).

How to enjoy A delightful, well-flavoured tomato for eating fresh or making great salsas.

Few seeds

Streaked red and yellow flesh

Black Krim

🌿 ☀ ☀ **Heirloom, 75 days** Also known as 'Czerno Krimski' or 'Black Crimean', this tomato comes from the south of Ukraine.

Characteristics The richly coloured fruits have a soft texture and an intense flavour, sometimes described as smoky. Fruits can be prone to cracking.

How to enjoy The large fruits are wonderful grilled and served with bacon for breakfast.

Rich mahogany-coloured skin

Cherokee Purple

🌿 ☀ **Heirloom, 80 days** Popular heritage variety distributed by John D. Green of Tennessee whose source said her ancestors obtained it from the local Cherokee people. Shows some blight tolerance.

Characteristics These maroon fruits stay green at the shoulders and vary from 150 to 400g (5 to 14oz). The flesh has a sweet intensity.

How to enjoy Eat fresh as you would an apple.

Plum

Plum tomatoes have an elongated, oblong shape, being typically 7–9cm (3–3½in) long and 4–5cm (1–1½in) in diameter. Use these for making sauces and purées – their flesh has a higher solid content than most other tomatoes, which gives great flavour and texture. They are particularly known for their use in canning. Mini-plum varieties have a thicker flesh than cherry tomatoes and are equally good for snacking. They are ideal for packing in lunchboxes, as they won't bruise easily.

Olivade

¶ ☼ ♔ **F1 hybrid, 72 days** Excellent, productive plants suitable for greenhouse or outdoors, yielding around 60 fruits per plant. Shows good disease resistance.

Characteristics Large plum fruits averaging 100g (4oz) with deep red skin and juicy flesh. A good long-keeping variety with few seeds.

How to enjoy In richly flavoured sauces to eat with pasta.

Black Plum

¶ ☼ ☼ **Heirloom, 82 days** This plant from Russia crops well even in cooler climates. Plants may be grown as semi-bushes.

Characteristics The firm-textured, purplish red fruits are an appealing even plum shape and are very juicy.

How to enjoy A crisp-textured tomato for snacking or good roasted in olive oil.

Deep red skins

Juicy flesh

Deeper colour at shoulders

San Marzano Lungo

🌱 ☀️ **F1 hybrid, 75 days** A popular strain of the well-known Italian plum tomato, which also has miniature and pink variants. Crops consistently over a long period.

Characteristics Elongated plum fruits with a dry, mealy texture.

How to enjoy Use in sauces, soups, and tomato paste; also excellent for drying.

Stubby ends

Roma

🌱 ☀️ **Heirloom, 78 days** Compact bush bred by the United States Department of Agriculture in 1955. Later selections have improved disease resistance.

Characteristics Looks like a traditional Italian plum tomato in shape and colour. Good taste with solid, meaty flesh.

How to enjoy Specifically bred for canning but suitable for most cooking purposes.

Rio Grande

🌱 ☀️ ☀️ ☀️ **Heirloom, 80 days** Productive semi-bush plants show good disease resistance and tolerance of extreme temperatures. Widely grown in Greece.

Characteristics Fairly blocky plum-shaped fruits, which are a good deep red colour.

How to enjoy Ideal for sauces and making fresh tomato juice. Grown commercially in Turkey for sun-dried tomatoes.

Pointed tip

Principe Borghese

♈ ☀ **Heirloom, 75 days** Italian heritage variety known by 1911. Though a bush plant it needs staking due to the weight of fruits produced.

Characteristics Heavy clusters of small plum-shaped fruits. A good long-keeping variety with few seeds.

How to enjoy Can be enjoyed in soups and sauces but best used for drying.

Few seeds

Alaskan Fancy

♈ ☀ ☀ **Heirloom, 55 days** An early and reliable cropper. It was bred for the cool climate and short growing season of Alaska. Performs well in poor summers.

Characteristics The medium-sized, juicy, red tomatoes have a good flavour compared to many other plum varieties.

How to enjoy Great for salads but flavour improves when cooked for sauces and soups.

Aviro

♈ ☀ ☀ **F1 hybrid, 70 days** Formerly known as 'Orange Plum', this productive plant shows good virus resistance.

Characteristics These medium-sized, pointed plums are orange-scarlet in colour and, like other orange varieties, have particularly high beta-carotene and vitamin C levels.

How to enjoy Pretty fruits to include in salads or excellent roasted.

Long Tom

🌿 ☀ **Heirloom, 85 days** 'Long Tom' originated in Pennsylvania but is closely related to the elongated Italian plums.

Characteristics These deep red plums have a good flavour and little juice.

How to enjoy The low acidity and firm texture make this variety excellent in sandwiches. Also a good choice for infusing flavour into sauces and soups.

Blondköpfchen

🌿 ☀ ☀ **Heirloom, 78 days** Translated as "little blonde girl", this very productive German plant yields huge trusses of long-lasting fruits.

Characteristics Small golden-yellow baby plum-shaped fruits are sweet and resistant to splitting.

How to enjoy One of the best mini-plum varieties for roasting whole.

Jersey Giant

🌿 ☀ **Heirloom, 85 days** Like the chicken of this name, the Jersey Giant tomato is from New Jersey in the USA.

Characteristics Large, elongated plum-shaped tomatoes with firm, sweet flesh and few seeds.

How to enjoy One of the best varieties for making tomato paste.

Particularly large fruits

Purple Russian

☀ ☀ ☀ Heirloom, 80 days High-yielding plants from the Ukraine.

Characteristics Violet-coloured plums averaging around 150g (5oz) with a firm texture and excellent flavour. They keep well.

How to enjoy Attractive sliced in salads or will dry well for winter use.

Many seed chambers

Speckled Roman

☀ ☀ Heirloom, 85 days The result of a cross between 'Banana Legs' and 'Antique Roman' that took place in the Illinois garden of John F. Swenson. It is similar to 'Casady's Folly' and 'Opalka'.

Characteristics Beautiful, pointed plums with stripes and speckles and a firm meaty flesh.

How to enjoy Enjoy as a work of art before halving and roasting.

Unusual, striped skin

Yellow Butterfly

☀ ☀ Heirloom, 68 days A prolific producer of mini-plum tomatoes similar to 'Ildi' (p73), this plant is becoming more widely grown. Rampant plants produce an extraordinary number of fruits in a truss.

Characteristics These ornamental dainty fruits have a pleasant sweet taste.

How to enjoy Cut trusses of tomatoes and hang indoors. Nibble at your leisure.

Plum- or pear-shaped fruits

Golden Sweet

🎋 ☀ 🏆 **F1 hybrid, 65 days** Mini-plum hybrid from Taiwan, which produces clusters of around 40 pretty yellow plum-cherry tomatoes.

Characteristics Tomatoes resemble oversized grapes and have a mild sweet flavour and firm texture. Resistant to splitting.

How to enjoy Ideal nibbling fruits for lunchboxes and picnics or use in sweet pickles.

Juliet

🎋 ☀ **F1 hybrid, 60 days** A mini-plum-shaped variety developed in Taiwan, sister to the well-known 'Santa' (below).

Characteristics A thick skin that is resistant to splitting makes 'Juliet' a very popular commercial tomato.

How to enjoy Lovely in Mediterranean-style casseroles but the thick skin means they are a little chewy when eaten fresh.

Deep red fruits

Yellow Santa

🎋 ☀ ☀ **F1 hybrid, 70 days** Bred by the Known-You Seed Company of Taiwan, the 'Santa' mini-plum variety is very popular with commercial growers. A number of similar varieties including this yellow form have now been released.

Characteristics Very sweet, glossy fruits with a tough skin, so they store well.

How to enjoy An ideal snacking tomato.

Rosada

🌱 ☀ **F1 hybrid, 75 days** A high-yielding plum-cherry hybrid from Taiwan that is sadly susceptible to splitting. One of the favourites in a 2007 RHS taste trial.

Characteristics The bright red mini-plums have a great depth of flavour.

How to enjoy These thin-skinned fruits are best eaten straight from the vine or used to flavour soups.

Floridity

🌱 ☀ **F1 hybrid, 62 days** These plants, from the English firm Tozer Seeds, produce many fruits over a long-cropping season. Voted top in an RHS taste test in 2007.

Characteristics These mini-plum-shaped tomatoes are resistant to splitting.

How to enjoy Ideal for nibbling straight from the vine and making soups and sauces.

Rudolph

🌱 ☀ **F1 hybrid, 68 days** Like its sister variety 'Santa' (p69), this was bred by the Known-You Seed Company of Taiwan. It produces long trusses of up to 25 fruits.

Characteristics Mid-sized plums with a firm texture and excellent flavour.

How to enjoy Halve and bake in the oven or a good choice for making soups or drying.

Firm fruits

Few seeds

Old Ivory Egg

🌱 ☀ ☀ **Heirloom, 75 days** This Australian variety, also known as 'Ivory Egg' and 'Australian Yellow Plum', was listed by the American Seed Savers Exchange in 1985.

Characteristics Sweet, citrussy, plum-shaped fruits are around the size of a small hen's egg.

How to enjoy Best eaten raw in salads, as the flesh tends to disintegrate on cooking.

Creamy yellow when fully ripe

Mini–Charm

🌱 ☀ **F1 hybrid, 75 days** Hybrid plant from the USA, which shows good resistance to disease. Vigorous plants with well-spaced-out leaves.

Characteristics These mini-plum-shaped, red fruits, just 7–10g (¼oz) each, are very juicy.

How to enjoy Tiny tomatoes with a very sweet flavour, they are ideal for lunchboxes.

Dasher

🌱 ☀ **F1 hybrid, 65 days** A mini-plum tomato grown commercially in Italy. Plants are sometimes offered to gardeners as grafted transplants.

Characteristics Appealing red mini-plums known for their excellent sweet flavour.

How to enjoy Straight off the vines as the ideal nibbling snack.

Orange Banana

🌱 ☼ ☀ **Heirloom, 80 days** A high-yielding Russian heirloom that can be grown under cover or outdoors.

Characteristics These deep orange fruits are plum shaped with a pointed nipple-like tip.

How to enjoy The richly flavoured fruits make a wonderful golden sauce or are good for drying to use in the winter.

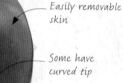

Often remains green at shoulders

Cornue des Andes

🌱 ☼ ☀ **Heirloom, 78 days** Collected in the Andes by a French seed collector, other names for this variety include 'Des Andes', 'Andine Cornue', and 'Poivron des Andes'. Plants can be prone to mildew.

Characteristics Elongated plum-, pimento-, or horn-shaped red fruits with pointed tips.

How to enjoy Their strong flavour is excellent in chutneys and ketchups.

Easily removable skin

Some have curved tip

Few seeds

Sunstream

🌱 ☼ **F1 hybrid, 72 days** Popular new hybrid from the Dutch firm Enza Zaden, which shows good disease resistance.

Characteristics Blocky fruits with glossy red skins have an excellent flavour.

How to enjoy Halve and roast these fruits with olive oil or use for drying.

Apero

† ☼ ♛ **F1 hybrid, 65 days** Winner of the tastiest tomato competition at West Dean, England in 2008, this British tomato is produced on many compact trusses.

Characteristics Glossy red fruits are renowned for their great flavour.

How to enjoy Pop them straight in your mouth to appreciate their succulent taste.

Chiquita

▽ ☼ **F1 hybrid, 63 days** Low-growing bushy plants are excellent for growing in pots and hanging baskets.

Characteristics These dusky pink baby plums have a firm flesh and superb flavour.

How to enjoy Add to a fresh fruit salad for an interesting dessert.

Dusky pink in colour

Ildi

† ☼ ♛ **Hybrid, 70 days** From the German firm Saatzucht Quedlingburg, 'Ildi' is a diminutive of the Hungarian girl's name Ildikó. Plants produce trusses of up to 80 fruits with a total of around 700 per plant.

Characteristics Mini-plum/pear-shaped fruits.

How to enjoy Hang trusses in a cool place and they will keep for weeks. The dainty fruits look pretty in an edible floral arrangement.

Weird and wonderful

Many tomatoes do not easily fit into any of the main categories. For example, there are a number of extreme variations of the plum shape – banana-, carrot-, sausage-, and flask-shaped fruits all appear. Pear-shaped variants with a distinct bottle neck are often considered a gourmet fruit. Oxheart varieties, which look like giant strawberries, often have a dense flesh and a delicious flavour. And blocky, squarish-shaped hollow fruits resembling bell peppers are ideal for stuffing.

Yellow Stuffer

☀ ☼ Heirloom, 85 days Also known as 'Gourmet Yellow Stuffer' or 'Yellow Cup', this variety was bred in the 1980s by Colen Wyatt of Petoseed in the USA.

Characteristics The deep yellow fruits are the shape of a bell pepper with thick walls and a hollow interior.

How to enjoy Stuff with a savoury filling of your choice, such a risotto or lentils.

Hollow cavities inside

Yellow Pear

☀ ☼ ☼ Heirloom, 75 days Also known as 'Beam's Yellow Pear', this is a very old variety that may date back as far as the seventeenth century. Vigorous plants produce fruits over a long season.

Characteristics These baby pear-shaped fruits with a sweet taste weigh around 15g (½oz).

How to enjoy Tasty fruits look very appealing in salads or to eat by the handful.

Distinct bottle neck

Red Fig

☀ ☀: Heirloom, 85 days This American variety from the eighteenth century is closely related to 'Red Pear'. It was traditionally dried for use in the winter as a fig substitute. The plant crops heavily.

Characteristics Small pear-shaped, juicy fruits.

How to enjoy Attractive fruits are good to eat fresh but would traditionally be sun-dried.

Japanese Black Trifele

☀ ☀ ☀: Heirloom, 80 days Despite its name, this variety, also known as 'Yaponskiy Trufel', is thought to be of Estonian origin.

Characteristics These fascinating fruits vary greatly but can be the size and shape of a large pear, richly coloured in red-maroon.

How to enjoy Eat cut into thick wedges to appreciate the wonderful rich flavour.

Split-resistant skins

Plum Lemon

☀ ☀: Heirloom, 81 days A vigorous, disease-resistant plant originating in St Petersburg, Russia. It was introduced to the USA in 1991 by Kent Whealy, who received it from a Moscow seedsman.

Characteristics The fruits usually are somewhat lemon-like in both shape and colour.

How to enjoy The firm, meaty tomatoes have a mild, vaguely citrus taste, best eaten fresh.

May ripen to a golden-yellow colour

Yellow Oxheart

☀ Heirloom, 80 days A family heirloom from Virginia that is thought to date back to around 1915. The finely divided leaves can show poor resistance to disease.

Characteristics The medium to large heart-shaped fruits are light yellow in colour and have good juicy flesh.

How to enjoy The succulence and excellent flavour of this tomato make it ideal for juicing.

Cuor di Bue

☀ Heirloom, 75 days Translated as "heart of beef", this Italian oxheart variety is similar to the French 'Coeur de Boeuf'. Lax plants produce very dense heavy fruits and need to be well supported.

Characteristics These tomatoes have rich red flesh and few seeds. Fruits usually weigh around 200g (7oz) but can exceed 400g (14oz).

How to enjoy Wonderful in simple salads.

Orange Strawberry

☀ Heirloom, 75 days Originating in 1993 in the garden of Marjorie Morris of Indiana as a chance seedling from 'Pineapple' (p53), this beautiful orange variety is sometimes late to mature.

Characteristics The strawberry-shaped fruits have a firm, dry flesh that is richly flavoured.

How to enjoy Admire the fruits as *objets d'art* before enjoying their strong taste.

Strawberry-shaped fruits

Green Bell Pepper

¶ ☼ **Heirloom, 75 days** This variety was bred by Thomas P. Wagner of Washington, USA, founder of Tater Mater Seeds, from his 'Brown Derby Mix'.

Characteristics Unusual green-and-yellow-striped, hollow tomato. The flesh is firm, very much like a bell pepper.

How to enjoy Ideal for stuffing with a savoury filling or slice into rings for salads.

Hollow inside

Reisetomate

¶ ☼ **Heirloom, 65 days** Also known as 'Pocketbook' and 'Voyage', this variety may be from Austria, but is similar to tomatoes used by native peoples in Central America. Translates from German as "travel tomato".

Characteristics These bizarre fruits resemble bunches of cherry tomatoes stuck together.

How to enjoy Traditionally said to have been eaten by travellers, one piece at a time.

Multilobed fruits, may have many segments

Green Sausage

∀ ☼ **Heirloom, 65 days** Originally raised in the USA by Thomas P. Wagner who called it 'Greensleeves', this variety was introduced commercially in 1998 by Kees Sahin of Sahin Zaden in The Netherlands.

Characteristics Elongated sausage-shaped fruits that are often curved like a banana.

How to enjoy With its distinctive appearance and taste, this is a tomato to savour fresh.

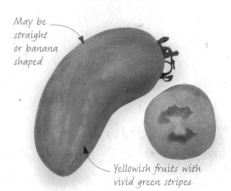

May be straight or banana shaped

Yellowish fruits with vivid green stripes

In the garden

Growing tomatoes in your garden gives you endless pleasure. Firstly, decide on where to grow your plants and whether you want to start from seeds or plantlets. The rituals of plant nurture – watering, feeding, removing sideshoots – are tasks to look forward to every day. Once you've harvested your first successful crop, you'll want to expand your tomato skills – try some grafting, embark on crop rotation, and even create your own variety!

A tomato for every climate

Because tomatoes originated in the coastal highlands of South America they are not well adapted to extremes of temperature. However, some plants are more suited to some climates than others.

Choosing appropriate tomato plants for your environment will give you a better chance of a successful crop. No tomato plants are able to survive frost, but there are a large number that have been developed to be able to crop successfully in cooler climates. However, if you have set your heart on growing a particular variety, you may be able to adapt the microclimate of your garden to suit it. For example, a variety that prefers warm conditions, such as 'Pink Accordion' (p55), may thrive in cool conditions if set against a sunny wall and sheltered from cold winds.

Strange as it may seem, many cold-tolerant plants are also best for growing in hot climates. That is because they are generally early-fruiting plants; if started at the beginning of the year, they produce a good crop before the intense heat of summer.

Survive the chill
For cold climates, try standard globe varieties 'Glacier', 'Polar Baby', 'Siberian' (p34), and 'Sub-Arctic Plenty' (p34); plum varieties 'Alaskan Fancy' and 'Aviro' (p66); and the cherry tomato 'Black Cherry'. The French 'Carmello' is very productive even in cool weather, as is 'Stupice' (p27). 'Scotland Yellow' (p28) is a good, reliable choice.

Black Cherry
(p38)

Glacier
(p23)

Alaskan Fancy
(p66)

Stand the heat

The Florida Agricultural Experiment Station has released many heat-tolerant tomato plants, starting with the popular variety 'Marglobe' in 1925. Others worth trying are standard globe varieties 'Mule Team' (p25), 'Stone', 'Thessaloniki', and 'Tropic'; the Italian beefsteaks 'Costoluto Genovese' (p56) and 'Costoluto Fiorentino', as well as 'Great White', 'Homestead', and the ribbed beefsteak 'Pink Accordion' (p55); and plum tomatoes 'Amish Paste' and 'Roma'.

Roma
(p65)

Legend
(p55)

Costoluto Fiorentino
(p57)

Come with a raincoat

The most important consideration in wet climates is choosing plants that resist fungal diseases, particularly the dreaded late blight (pp114–19). This is caused by the fungus *Phytophthora infestans* and can wipe out an entire crop in wet summers. Large-fruited plants tend to be more susceptible to the disease. Sadly, no tomato plant can be said to be totally blight-immune, but resistant plants worth trying include 'Early Cascade', 'Fantasio', 'Ferline', and 'Legend.

Ferline
(p23)

Where shall I grow my tomatoes?

Tomatoes are very adaptable plants and most will grow well in containers, in the ground, as well as under cover. Choose a site where you can appreciate the plants when they come into their full bloom.

Patio pot

Growing tomatoes in pots means they are easy to transport so you can position them where you like. They prefer a warm, sunny, sheltered position where they will have protection from excessive wind and rain.

Large plants may become too top-heavy for the pot or exhaust the water and nutrient supply in the compost too quickly. The smaller bush or dwarf plants are the best choice because they will not impinge on any seating space. Dwarf plants such as 'Balconi Red' (p44), 'Minibel', and 'Micro-Tom' (p43) should crop successfully even in a 15cm (6in) pot. However, for most other plants it is a case of bigger is better when it comes to pot size. Terracotta pots look attractive but dry out quickly, alternatively use glazed (or plastic) pots.

Hanging basket

Cherry tomatoes cascading from a hanging basket can be very decorative, as well as productive. Bush plants, such as 'Tumbler' (p36) and 'Tumbling Tom Red' (p49), are easiest to use: they do not need pinching out and naturally trail. Hanging baskets can dry out very quickly, particularly in warm breezes, and may need watering several times a day. Line the basket with perforated black plastic, use a basket with an integral water reservoir, or put a handful of ice cubes to melt into the basket each morning.

Alternatively, why not grow your tomatoes in an upside down planting pouch? Plant a dwarf, bush, or trailing tomato plant into these compost-filled plastic pouches, hang from the centre of a greenhouse or a tree, and watch the colourful cascade emerge.

Growing bag

These sealed plastic bags provide a very popular and fairly cheap method of growing tomatoes, particularly cordons plants such as 'Ailsa Craig' (p32) or 'Gardener's Delight' (p42). They are filled with a proprietary growing medium, usually based on fertilized peat or a peat substitute, but you could try using ordinary bags of compost. Many, however, contain just 35 litres (61.6 pints) of compost and are very shallow, giving little space for healthy root growth. They can be quite tricky to keep evenly watered and regular feeding is essential. You could stand bottomless pots on top of the growing bag and fill these with compost so that the plant can root to a greater depth (this is called ring culture).

Basket case Position your hanging baskets in a porch so that you can pinch a few tomatoes as you enter or leave the house.

In the bag Cordon tomatoes are ideal for growing bags. Tie cane poles to greenhouse struts or a trellis to hold them upright.

In the ground

Any type of tomato plant may be grown in open ground. This allows the plants to utilize available water and nutrients from the soil. The plants may also be planted deeply, which encourages them to produce plenty of fibrous, feeding roots. However, depending on the weather and soil fertility, they may still need regular watering and supplementary feeding. In warm climates or in particularly good summers in temperate regions, tomatoes grown outdoors usually produce the most flavourful fruit. It doesn't matter whether the plants are crammed into a flower border (cottage-garden style), in the vegetable garden, or perhaps on an allotment or in a field. Good crops can result from plants in any of these situations. It is, however, important to consider choosing plants that are more resistant to pests and diseases (pp114–19); tomato blight in particular can be a problem with outdoor crops. Some outdoor bush tomatoes may be allowed to sprawl over the ground, but this makes them vulnerable to pests such as slugs and to rotting. Support them with canes and twine or proprietary support systems.

Lean on me Cordon plants need a sturdy support, such as a single stake. You could use horizontal training wires on a warm wall.

Formal potager

Originating in France, potagers are kitchen gardens of fruit and vegetables laid out in decorative patterns, often with geometric beds edged with low box hedges. They usually include flowers and herbs, which are planted among the vegetables to encourage pollinating insects, and for decorative effect. Tomato plants, with their vines of colourful fruits, are great for such gardens. While bush plants tend to be the most popular for outdoor use, they usually produce all their fruits at once, which is not ideal if you want the potager to look decorative for a long season. It may be better to grow cordon plants trained to sturdy stakes. To maximize visual appeal, train the vines up cast-iron obelisks or over garden arches so that the trusses of ripening tomatoes can hang down and catch the light.

Outdoor kitchen Growing tomatoes with other plants makes an attractive display.

Greenhouses and polytunnels

Growing tomatoes under cover is especially beneficial in climates that have cool, wet summers as it extends the season over which they bear fruit and offers protection from diseases, especially the fungal disease late blight (pp114–19). However, some pests such as whitefly also enjoy life under cover and can be more of a problem in a greenhouse than outdoors.

Polytunnels are usually much cheaper than greenhouses and consist of plastic film stretched over galvanized steel arches to form a walk-in tunnel. They are more susceptible to wind than greenhouses, so must be well braced and the arches secured. The curved shape encloses a larger volume of air than a greenhouse, so a polytunnel stays warmer for longer. Plastic films usually contain ultraviolet light inhibitors and also reduce risk of mildew, rot, and insect attack.

An unheated greenhouse or polytunnel does not give frost protection in temperate regions, but extends the season. Tomato plants will crop all year round with supplementary heating. The limiting factor for plant growth in winter is often lack of light. Orientate the greenhouse east to west to increase light transmission, and clean the glass: clear panes let through nearly 50 per cent more light than dirty ones.

Growing tomatoes under cover

You can plant tomatoes directly in soil beds in the greenhouse to give them a greater rooting area. However, growing them in the same beds every year exposes the plants to many soil-borne pests and diseases. Use growing bags or containers to avoid this.

Poor ventilation can lead to very high temperatures on sunny days as well as inadequate air flow, causing high humidity, which may encourage fungal diseases. Automatic window openers and vents are worth investing in. Apply shading to the glass in summer. On hot days, water the floor of the greenhouse to cool the air, but without wetting the tomato foliage.

Window cleaning Keep the glass of your greenhouse clean to allow maximum light transmission to your plants.

Warm up Polytunnels encase more air so stay warm for longer then greenhouses. >

Preparing the soil

Tomatoes are greedy feeders and grow best in a rich, fertile, and moisture-retentive soil. Digging over the soil and incorporating plenty of organic matter will get your plants off to a good start.

Do-it-yourself compost

The best way to improve and enrich soil for tomatoes is to add plenty of organic matter and home-made compost is an excellent source. Composting utilizes the biochemical process of decomposition, carried out by naturally occurring organisms. Using a compost bin will facilitate the process, but you could build a traditional open heap in a corner of the garden. Open heaps are,

however, more likely to have significant weed populations and the resulting compost may contain fewer nutrients because they will have been leached out by rainfall.

A compost heap needs air, moisture, and warmth to keep active, so keep it fairly sheltered, just moist, and aerated. Regularly turning a heap increases oxygen levels and accelerates the process, but is not essential.

MAKING YOUR OWN COMPOST – convert waste into valuable soil conditioner

1 Collect uncooked vegetable waste from the kitchen, and other soft, green matter from the garden, to form the basis of compost.

2 Add the green waste to the bin with drier, fibrous material such as shredded prunings, dead leaves, paper, and cardboard.

Enrich the soil

Late winter or early spring is the best time to prepare the soil for growing tomatoes. As well as home-made compost, you can use other organic materials such as well-rotted manure or composted bark. Clear the bed of any weeds first. If you don't want to disturb the soil structure, you can simply spread a 7.5cm (3in) layer of organic matter on the surface of the soil and let the worms take it in. However, if your home-made compost contains a lot of weed seeds, it is better to dig it in to stop the seeds germinating.

Garden compost makes a great soil conditioner. Use it as potting compost too, if it is reasonably free of weed seeds.

Seed or plantlet?

Watching a seed transform into a plant is a great pleasure. Seeds are cheap and many varieties are readily available. Alternatively, you can cut out the germination period and buy a healthy young plantlet.

Starting with seeds

Mail-order catalogues and websites offer a much greater choice of seeds than garden centres. Any reputable supplier will recommend suitable varieties for your local conditions (pp80–81); check also that the variety's growth habit is suitable for you. The cost of seeds may vary so check the approximate seed count on the packet as well as the price. Seed of an F1 hybrid is more expensive than open-pollinated seed.

It can be satisfying to buy seeds from small family suppliers, but you should be aware that if they grow a lot of varieties in a small area there is a greater chance of cross-pollination and therefore variation in the plants. If you are interested in heirloom varieties, it is worth joining one of the non-profit organizations dedicated to the preservation of heirloom seeds. (See p187 for useful websites.)

CHOOSE YOUR SEEDS – each one a potential plant

Naked seeds Most seeds are supplied as dried seeds; store in a cool, dry place out of the light and they will keep for a long time.

Pelleted seeds A clay coating makes these seeds easy to handle. Germination rates are less erratic but they don't keep as long.

Starting with plantlets

Although it is less expensive to raise your own plants from seeds, buying in seedling plants, or plantlets will save you time that you would otherwise spend raising them yourself. Select good, sturdy plants with deep green leaves that are free from spots and bugs. Good nursery staff should not be upset if they see you checking the undersides of the leaves for insects or tipping a sample plant from its pot to check for vigorous, healthy roots.

If plantlets are particularly small and have undeveloped roots, pot them into 10cm (4in) pots and grow on until the roots fill the new pots. When conditions allow, plant out any that have flowers (pp98–101).

Tomato plantlets
Buy plantlets in individual modules or biodegradable pots, so you can pot on or plant them out without any root disturbance.

How to sow seeds

Sowing tomatoes is not a precise science, but for gardeners in temperate regions it's best to start them 6–8 weeks before the usual date of the last frosts, to give them a long growing season.

Sowing seed too early can result in lanky seedlings, if there is insufficient light early in the season. How much seed you sow depends on how many plants you require. If you need just a few plants, sow up to three seeds in a 7.5–10cm (3–4in) pot. For larger quantities, sow seed in individual modules or seed trays. It is best to use a specifically designed seed compost because a final potting mix may have too high a level of nutrients, which can actually inhibit germination. Seeds usually germinate after about five days at 18–24°C (64–73°F), but some older varieties may take 2–3 weeks. Make sure that the compost does not dry out during this time by covering the container with a plastic bag or placing it in a heated propagator with a lid.

SOWING TOMATO SEEDS INDOORS – sow early for a longer season

1 Fill a clean container with seed compost so that it is slightly overfull. Level the surface by drawing a piece of wood across the rim.

2 Lightly firm the compost with a block of wood or the back of your hand to create an even sowing surface.

3 Water well with a fine-rosed watering can; alternatively, stand the container in a tray of water until the surface is moist. Drain.

4 Sow the seeds evenly over the surface. It is easier to space the seeds if you trickle them from the palm of your hand.

5 Cover the seeds with a thin layer of compost, firm, and label. Place in a warm position, such as a windowsill or heated propagator.

6 Once germination begins, remove cover, if used, to decrease humidity and grow on the seedlings at 16–20°C (61–68°F).

how to sow seeds   **93**

The first few weeks ...

Watch your seedlings as they begin to grow and the first few leaves appear. Seedlings will be competing for the same water, nutrients, and space to grow so will need separating into other containers.

The first two leaves to appear are simple, oblong-shaped seed leaves. Once the seedlings have their first pair of true leaves (p12), they can be thinned or pricked out. If you have sown two or three seeds in individual pots or modules and you only want one plant, you need to thin them out (below).

Seed sown thickly in pots or seed trays is at greater risk of fungal infection, so seedlings should be pricked out into individual pots (opposite) as soon as they are large enough to handle. Use clean, 7.5cm (3in) pots filled with moist compost.

Grow on the pricked-out seedlings indoors in well-ventilated conditions in a bright spot, but not in direct sun, at 16–24°C (61–75°F). They will cope with slightly cooler temperatures overnight. Keep the compost moist, not soggy. There is no need to feed the seedlings at this stage.

THINNING OUT SEEDLINGS – weeding out the weakest

Choose the strongest seedling in each pot or module and remove the others. Simply pull them out gently or, to avoid disturbing the roots of the strongest one, nip them out at soil level with clean scissors or fingernails.

PRICKING OUT SEEDLINGS – handle them with care

1 Hold each seedling by its leaves (the fragile stems are easily damaged) and use a dibber or sturdy plant label to ease the seedling's roots gently from the compost.

2 Plant each seedling in a pot of fresh compost, filling in carefully so the seedling is at the same level as before, or slightly deeper if the stem is a bit lanky. Firm gently around the stem and water in well.

Alternatively, make a hole in the centre of a filled pot with a pencil or your finger, and lower the seedling roots into the hole. Firm.

Toughening up

Young tomato plants may be planted outside once all danger of frost is past, but those started off in a protected environment will need a transition period to get them used to the conditions outdoors.

Hardening off is the process of allowing young plants to adapt to the wind, sunlight, and fluctuating temperatures outdoors. Put the plants outdoors for an hour on the first day, two hours on the next, and work up to five or six hours. Avoid exposing them to full sun or wind.

Once plants are hardened off, they can be planted outdoors in suitably prepared soil. Be ready to protect young plants with a layer of horticultural fleece or newspapers in the event of an unexpected cold snap.

Hardiness, or ability to withstand cold, is a complex quality in plants. No tomato plants are able to withstand frost, but their ability to cope with cool conditions varies. Some plants such as 'Sub-Arctic Plenty' (p34) can set fruit under quite cool conditions, but others will need shelter to create a favourable microclimate.

Half-way house

You can use a cold frame or cloche as a half-way haven for hardening off young tomato plants, between indoors and out.

In areas with a short growing season, you can also use cold frames and cloches to protect young tomatoes after planting and help extend the season. Make your own (below) or use ready-made models. Another way to shelter a young plant and enable you to plant it out a month or so earlier is a wall o'water – a circle of water-filled plastic bottles around the plant. The water absorbs heat during the day and release it at night.

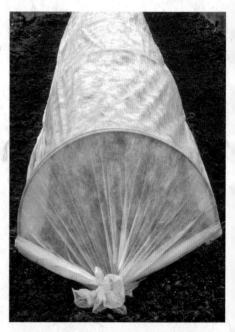

Tunnel cloche To make a cloche, drape a horticultural fleece over wire hoops. It will protect plants in the period after planting.

Cold frame Insulate your plants so they
slowly acclimatize to cool temperatures. Open
the frame at intervals for good ventilation.

Planting

Plant tomato plants into soil beds or containers when they are
15–30cm (6–12in) tall, with well-developed, healthy rootballs.
Moist (not soggy) rootballs slip out of pots more easily.

Soil beds

When planting out in outdoor beds, use an
appropriate support system – try wire cages
for bushes and wooden stakes for cordons.
Dwarf plants don't need a support. Choose
an overcast, still day to reduce stress to
the plant from sun and wind and plant in
a sheltered spot, such as against a sunny
wall or downwind of a hedge or trellis.

PLANT IN A SOIL BED – plant deeply for healthy roots

1 While soaking the plant pot in water or a
dilute seaweed solution (for 2–3 hours), dig
a planting hole, about twice the pot's depth.

2 Hammer in a sturdy stake next to the
planting hole. Knock the plant from its pot
and place it into the soil.

3 Backfill with soil so that the stem is covered at least up to the level of the seed leaves. This will encourage new roots to form, which stabilizes the plant and can increase yield, as there are more feeding roots. Firm.

4 Water in well and label. In areas where cutworms are a problem and may nibble the stem, make a newspaper or cardboard collar and slip it around the base of the stem. This should protect the young plant until it is robust enough to resist attack.

Containers

When you are transferring young tomato plants into containers such as pots, hanging baskets, or growing bags, you need to observe the same basic principles as for planting out (pp98–99).

One advantage of containers is that you can start the tomatoes earlier and keep the planted-up containers under cover until conditions are favourable outdoors, which will give you an earlier crop. You can't do this with growing bags, however, as they are difficult to move once planted.

You may wish to plant a trailing or dwarf tomato plant into a hanging basket. Keep it in a light, airy room, or in the greenhouse, at a minimum of 15°C (59°F), until all risk of frost is past.

Tomato plants in pots and growing bags need a sturdy support system of canes and twine, stakes, or wires on fencing or walls. Bamboo canes are often used, but they may not be able to cope with the weight of some of the more vigorous plants. Opt for stakes in this instance. You can help to hold canes in growing bags upright by tying the top of each cane to a greenhouse strut or some other strong support outdoors such as a trellis.

PLANTING INTO A GROWING BAG – position the bag before planting it

1 Cut 3 planting holes in the plastic. Add a watering funnel (made from a baseless plastic bottle) to avoid water run-off.

2 To plant up the bag, water each plant well and insert it in its hole at the same depth or slightly deeper than before. Firm.

Protecting your plants

When positioning containers, choose a spot that provides shelter from strong winds and excessive rain. The most effective barriers are not solid screens such as walls or fences, but those that are around 60 per cent permeable, which filter the wind rather than causing damaging eddies. However, in cool areas, tomatoes trained against a sunny wall will benefit from reflected heat.

You can also improve drainage on heavy soils by building raised beds or adding plenty of organic matter to increase the moisture retention of light soils.

Create an effective screen from drying winds by growing climbing plants such as runner beans up a simple trellis screen. >

3 Punch some drainage holes into the base of the bag, to avoid the compost becoming stagnant. Then water the plants thoroughly.

4 As the plants grow, tie in the main stems to the supports, using soft twine in a figure-of-eight – don't tie the stems too tightly.

Crop rotation

If you grow tomatoes as part of a larger vegetable plot, you might want to include them in a crop rotation plan, along with the other vegetable crops, to protect them from pests and diseases.

The basics

Growing the same crops year after year in the same soil can lead to a build-up of soil-borne pests and diseases. If you rotate annual crops around the garden so that they are not grown in the same area again for at least three years, you should avoid such a build-up. Vegetables are usually grouped together for crop rotation according to their varying nutrient and cultivation needs and moved around in sequence.

The practice of crop rotation has other benefits. It helps to make efficient use of the soil nutrients. Also, alternating between deep-rooted crops and more fibrous-rooted ones can help to improve the soil structure. However, the rotation needs to be carried out over a much longer time scale than the usual three or four years to be completely effective, and keeping to a strict rotation in a small garden may be impractical.

Keep on moving

To follow a four-year crop rotation, divide the vegetable garden into four areas. On the first plot, grow tomatoes and other members of the same (*Solanaceae*) family, such as aubergines and potatoes. You can also include with this group root crops, such as carrots and beetroot. In the second year, move the tomatoes and roots to the second

plot, and in the first plot grow members of the onion family. In the third year, move the tomatoes and roots to the third plot, the onions to the second plot, and grow legumes (peas and beans) in the first plot. Most vegetables in the legume family have nodules on their roots; these house bacteria that can fix, or store, nitrogen from the soil, making it available for the following crop.

It is traditional to follow legumes with nitrogen-hungry brassicas, such as cabbages and broccoli, before using the same ground for tomatoes and root crops once more. If possible, grow tomatoes and potatoes apart – they share many of the same diseases.

Fallow seasons

Keeping soil fallow – growing nothing on it for a season – will reduce populations of some soil-borne pests, such as root-knot nematodes (microscopic worms); they will die out through lack of food. Dig over the soil to expose the nematodes and other pests to the drying sun and to feeding birds. This will refresh the soil ready for growing next year.

A four-year cycle The photographs opposite show 4 plots of land. Follow the plots to find out what you should grow in each, every year in a 4-year crop rotation. >

Plot One
Year 1 Tomato family **Year 2** Onion family
Year 3 Legume family **Year 4** Brassica family

Plot Two
Year 1 Brassica family **Year 2** Tomato family
Year 3 Onion family **Year 4** Legume family

Plot Three
Year 1 Legume family **Year 2** Brassica family
Year 3 Tomato family **Year 4** Onion family

Plot Four
Year 1 Onion family **Year 2** Legume family
Year 3 Brassica family **Year 4** Tomato family

A little help from friends

The practice of growing plants that may have a beneficial effect on their neighbours is called companion planting. Plants that repel pests or that attract pollinating insects make good companions for crops.

Keep companion plants 30–60cm (12–24in) from the main stems of tomato plants – close enough to be beneficial, without competing with the tomatoes for nutrients. In a greenhouse or polytunnel, place pots of companions among the crop.

Good companions

Ideal plants to grow with tomatoes include basil, chives, onions, carrots, mint, and parsley. Marigolds, especially the French marigold (*Tagetes patula*), are thought to be particularly good at repelling whitefly (pp116–17) and even nematode pests that may be present in the soil. Marigolds are long-flowering plants and, like nasturtiums (*Tropaeolum majus*), will attract many pollinators. The beautiful, blue-flowered herb, borage (*Borago officinalis*), is said to deter moths that may damage tomato plants (pp116–17).

In greenhouses prone to infestations of whitefly or fungus gnats, it is worth growing a few plants of the common unicorn plant (*Proboscidea louisianica*), also known as the common devil's claw or elephant's tusk. This pretty plant with showy gloxinia-like flowers has oval, wavy-edged leaves, which are covered in glandular hairs that sparkle in the light. The leaves are extremely sticky and act as flypaper, catching many small insects. The plant is considered a noxious weed in some American states because the strange, clawed seedpods can damage the eyes of livestock. Allow it only to set sufficient seed to grow a few new plants for the next year.

Friendly fungi

Mycorrhizae are soil-borne fungi that form symbiotic (mutually beneficial) relationships with plants. They colonize the roots of a nearby plant, such as a tomato, and extend their network (mycelium) of fine, sticky filaments, known as hyphae, into the nearby soil. A mycelium increases the surface area of the roots of a host plant, enabling the host to absorb more food and water.

It is also thought that the sticky hyphae protect the plant from soil-borne fungal diseases, making it difficult for them to invade the host. Mycorrhizae thrive in rich, organic soils, but are destroyed by digging and the high salt content of chemical fertilizers, so organic cultivation methods are thought to be particularly beneficial.

Best friends (clockwise from top left) mint (*Mentha*), chives (*Allium schoenoprasum*), French marigolds, and nasturtium. >

Mulching

A mulch is a protective covering of the soil's surface. It's a good idea to apply a mulch just after you have planted out to warm the soil and minimize weed growth. Choose from inorganic and organic mulches.

Mulches reduce water loss from the soil, insulate it from excessive cold and heat, and deter weeds. Inorganic mulches, such as black plastic sheets, gravel, or woven membranes, control weeds most effectively. Black mulches absorb heat and warm spring soils, but may overheat soils in hot temperatures. Red plastic mulches reflect intense red light to young plants and may increase crops. Most mulches are applied at planting, but light-coloured straw reflects the sun away from the soil so is better applied once plants are growing strongly. Other organic mulches include dried glass clippings, well-rotted animal manure or bark, wilted comfrey leaves, straw, or layers of newspaper. Mulches should not contain any residual weedkiller, which could harm the tomato plant and crop.

USING INORGANIC MULCH – stop the weeds with a plastic sheet!

1 Lay weed-suppressing membrane over the soil, then cut flaps in it and knock in a stake. Replace the flaps after planting.

2 Tie the plant to the stake with string and flatten the membrane firmly around the base of the plant.

USING ORGANIC MULCH – these attractive mulches allow water to permeate

Spread loose organic mulches, such as layers of grass clippings, to a depth of 7.5cm (3in). You can also top them up later on as the plant grows. This will stop the soil drying out and stressing the plant roots.

Food and drink

Like people, plants cannot exist without water so make sure your tomato plants have enough to drink. Food is important too and can boost your tomato crop. But with both food and drink, balance is key.

The water of life

Plants need water for transpiration, where moisture evaporates from open pores on their leaves, drawing more water up from their roots through tiny vessels. This process cools the plant and creates a transport system carrying water and nutrients around the plant. Transpiration is greatest in a dry, warm atmosphere and substantially increases in windy conditions. In reasonably sunny conditions, a typical tomato plant requires around 0.75 litres (1½ pints) of water per day. This varies greatly, depending on the type and state of the soil or compost and how well it drains, the stage of growth of the plant, and the weather. Container-grown plants need plenty of water because their roots cannot spread far. Feel the soil below the surface with your finger to see if it needs water.

Water with care Roots need a moist, not wet, growing medium. Soggy soils can be low in oxygen levels, leading to root death.

Split skin Irregular watering may lead to fruit splitting (pp118–19). Excessive watering may also produce large but tasteless fruits.

To feed or not to feed

Plants grown in open, well-nourished soil usually yield a good tomato crop without additional feeding. For the best crop from plants in pots or growing bags, feed them regularly after the first truss of fruit has set. Nitrogen (N), phosphorus (P), potassium (K), and calcium (Ca) are the most important minerals in tomato fertilizer.

Most home-made plant feeds smell dreadful, but are a cheap way of feeding your plants organically. Fill a container with comfrey leaves, nettles, or other fresh green weeds and pour boiling water over them. Cover and steep for a month, then strain the liquid and use well diluted in water. You could also use the liquid by-product from a wormery.

Too much fertilizer can be harmful to plants, so always follow the manufacturer's recommended dosages. Overfed plants may make excessive, soft growth, which will be vulnerable to disease and may delay flowering. A build-up of chemical salts in compost can also damage root growth. Never use lawn fertilizer on tomatoes: this is high in nitrogen, which will encourage lots of leafy growth rather than fruits.

Using a fertilizer
Specifically formulated fertilizer for tomatoes will ensure that your plants receive the right balance of nutrients.

Removing sideshoots

Making sure you prune sideshoots from your tomato plant regularly will ensure that it stays healthy. Keep your cuttings and use them to make new plants to increase your own crop or give them to friends.

Removing sideshoots produces a single stem, or cordon, which makes your plants easier to support and admits more air and water to their fruits. There is no need to remove sideshoots from bush or trailing types. The sideshoots form in every leaf axil – pinch them out regularly as soon as they can be handled. Cut off larger sideshoots at their bases to avoid tearing the main stem. In very hot climates allow each sideshoot to produce one leaf. This will protect the fruits from potential sun damage. When the cordon reaches the top of its support, pinch out the growing tips to encourage the fruits to ripen; also pinch out any flowers towards the end of the season. To create new plants, take side shoots early in the season, so that the new plants have long enough to ripen their fruits.

CUTTINGS ARE EASY! – recycle sideshoots into new plants

1 **Choose a sideshoot** that is reasonably firm at its base and can be handled easily, and snap it cleanly from the main stem.

2 **Place the sideshoot** in a jar or glass of clean water. Brown glass aids speedy rooting, but roots should appear anyway within a week.

3 **Pot up the sideshoot** carefully, once its roots are at least 1–2cm (½–¾in) long, in a pot of cuttings compost.

4 **Firm gently** to avoid damaging the fragile roots and water in. Grow on the cutting in the same way as a new seedling (pp94–5).

Large sideshoots of 30cm (12in) or more will root successfully, but keep them shaded or they may wilt.

Pollination

For a tomato plant to form a fruit, the stigma, or female part of a flower, must be fertilized by viable pollen. This will occur naturally with assistance from insects and wind but you too can help ...

Cultivated tomato plants are self-fertile – they can fertilize flowers with their own pollen. When the flower opens, pollen falls from the anthers, or male parts of the flower, onto the stigma. If the atmosphere is too dry, pollen will not stick to the stigma; in very wet conditions, pollen is not released from the anthers.

The optimal temperature for pollination is 18.5–26.5°C (65–80°F). In very cold conditions very little fruit will set, whereas very hot conditions kill the pollen. As there are fewer insects to pollinate flowers under cover, you can help yourself. Wait until there are many trusses of ripe flowers and carry it out at noon, when pollen is most abundant.

POLLINATING TOMATO FLOWERS – play Cupid to guarantee a good crop

1 The female part of the tomato flower (stigma) in most plants lies within a cone of pollen-bearing (male) anthers (bottom left).

2 Gently shake the flower clusters. Tapping the plant's wire or cane supports will also encourage the pollen to be released.

Busy bees Insects, such as the bumblebee (*Bombus terrestris*), play a vital role in pollination. As they collect nectar, they fly from plant to plant and distribute pollen. Many commercial units utilize this by installing colonies of bumblebees in cardboard hives among the tomato crops.

Pests and diseases

Healthy plants are better able to cope with pests and diseases and produce good crops, so it is important to identify and control any problems. The first step is to use good cultivation techniques.

Choose disease-resistant plants

Most modern F1 hybrid tomato plants are relatively disease resistant, as commercial plant breeders work to improve resistance to a number of fungal and other pathogens. Verticillium and fusarium wilts, for example, are soil-borne diseases that cause yellowing of the leaves, wilting, and premature death of plants. Once they build up in the soil, the only practical control is the use of resistant varieties, such as 'Roma' (p65), which are often designated by the letters "VF" in seed catalogues. "VFN" indicates a variety, such as 'Lemon Boy' (p35), that is also resistant to root-knot nematodes, worm-like pests that live in the soil. There are few blight-tolerant plants, although 'Ferline' (p23) and 'Legend' (p55) are particularly disease resistant.

Prevention is better than cure

Compacted, poorly drained soil stresses the plants, so prepare the soil well (pp88–89). Practise crop rotation (pp102–103) to avoid growing tomatoes after crops in the same family, which are prone to the same diseases. Eradicate weeds, particularly those in the nightshade family, which may harbour disease organisms. Maintain good plant hygiene by removing and burning all infected plant material – do not put infected material on a compost heap because some pests and diseases may survive and infect subsequent crops. Space plants to encourage air circulation and reduce spread of pests or diseases. Staking and removing sideshoots (pp110–11) promote air flow around plants. Try to keep the foliage dry to minimize fungal infections. Position plants where they benefit from the morning sun to dry out any dew from the leaves. Avoid watering the plants from overhead, particularly if it is late in the day.

A note about late blight

Late blight is the most common and most deadly problem that may affect your crop. It can have a devastating affect particularly in cool, wet climates. Do not leave unharvested potatoes in the soil, as they can be a source of infection to subsequent crops. If buying plantlets, check carefully to ensure you buy healthy stock. Growing tomato plants under cover will provide some protection from wind-blown spores. Use the charts on the following pages if your plant is showing signs of ill health.

Common problems with tomato plants include (clockwise from top left): early blight, late blight, blossom end rot, and sapfeeding whitefly. >

What's wrong with my leaves?

You can often distinguish a healthy plant from an unhealthy plant by its leaves, as they are usually the first to show symptoms of pests and diseases. Sometimes changes in leaf appearance may be a result of changes in temperature and are no cause for concern.

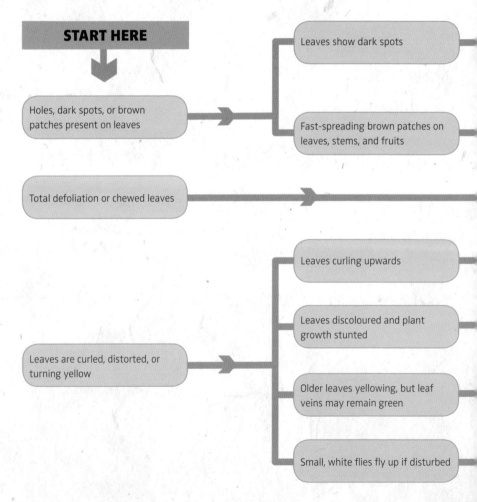

START HERE

Holes, dark spots, or brown patches present on leaves

Leaves show dark spots

Fast-spreading brown patches on leaves, stems, and fruits

Total defoliation or chewed leaves

Leaves curling upwards

Leaves discoloured and plant growth stunted

Leaves are curled, distorted, or turning yellow

Older leaves yellowing, but leaf veins may remain green

Small, white flies fly up if disturbed

CAUSE and TREATMENT

Spots often have concentric rings with a faded yellowing around them, particularly on lower leaves

→ **Early blight** (*Alternaria* species) is a fungal infection – Remove and burn all affected leaves immediately

Spots often have a thin circle of bright yellow surrounding them

→ **Bacterial leaf spot** – Remove all affected leaves and avoid wetting other foliage during watering

→ **Late blight** (*Phytophthora infestans*) – Burn affected plants; use copper-based fungicide as preventative spray

→ **Slugs, snails, or the caterpillars of various moth species** – Pick off pests

→ May be sign of **leaf roll virus**, but usually a result of cold nights – Will not affect the plant's growth

→ Viruses such as **tomato mosaic virus** or **curly top virus** – Burn affected plants and control aphids

→ **Magnesium deficiency** – Apply magnesium sulphate (Epsom salts) as a dilute solution to plant leaves

→ **Sapfeeding whitefly** (*Trialeurodes vaporariorum*) – Use sticky traps or biological control (*Encarsia formosa*)

what's wrong with my leaves? **117**

What's wrong with my tomatoes?

Keep an eye on developing fruits so that you can take action as soon as any symptoms of pests or diseases appear to avoid more fruits being affected. Some problems with your tomatoes can be resolved by simply watering them more regularly.

START HERE

Dark or pale colouring on fruits

- Dark or pale rings, or lesions present on leaves and stalk
- Blossom end of fruits is flattened and black
- Persistent pale rings appear on unripe fruits

Green or white hard tissue on fruits

- White, leathery areas, particularly on green fruits
- Partial or complete ring of unripe tissue at stalk end

Fruits distorted

- Skin of fruits split
- Distorted fruits with swollen areas at blossom end

Dark, concentric rings on stalk ends of fruits

Early blight (*Alternaria* species) is a fungal infection – Remove and burn all affected leaves immediately

Lesions on fruits are olive-brown coloured

Late blight (*Phytophthora infestans*) – Burn affected plants; use copper-based fungicide as preventative spray

Blossom end rot – lack of calcium caused by irregular watering; Control watering; apply feed of soluble calcium

Ghost spots caused by fungus (*Botrytis cinerea*) – Improve air circulation

Sunscald – Provide shading for plants

Greenback – usually caused by heat injury; Apply shading or choose resistant plants

Fruit splitting – caused by irregular watering and extremes of temperature; Water regularly

Catfacing – caused by poor pollination; Apply shade and damp down green-house if hot, put in sunny area if cold

Reaping the harvest

For the home gardener, the reward of a succulent, sun-ripened tomato that is rich in flavour is the ultimate goal, and so the fruits should be left on the plant to ripen to the peak of perfection.

The ripe stuff

The process of tomato ripening is governed by the temperature – a range of 18–24°C (65–75°F) is best for most varieties. As ripening gets under way, the green pigment (chlorophyll) breaks down and the orange-yellow (beta-carotene) and red (lycopene) pigments increase. It is the relative concentrations of the latter pigments that determine the colour of a ripe tomato.

Take your pick

Pick fruits as soon as they ripen so that the plant keeps producing new tomatoes. This is particularly important towards the end of the season, when you want the plant to concentrate on swelling the remaining fruits.

You can also help along the ripening process later in the season by removing any yellowing leaves from the bases of plants, to let in more sunlight to the tomatoes.

PICKING TOMATOES – find the ripest, juiciest tomatoes, pick, and enjoy

Small fruits Rather than picking lots of small fruits, it is easier to cut off the entire truss, with secateurs or a clean, sharp garden knife.

Large fruits Apply slight pressure with your thumb where stem and calyx join (abscission layer, or knuckle). Gently twist and break off.

Kitchen marvels The reward of spending just a few hours a week in your garden. Turn to page 132 to see how to enjoy them.

Beat the autumn frosts

Try to harvest all tomato fruits before the first frost. They may continue to ripen indoors (below), if picked immediately after a light frost, but a heavy frost will damage the internal tissues; use any frost-damaged tomatoes straightaway.

While the first frosts are light, you may untie a cordon plant from its support, lay it flat on the ground, and cover it with a layer of horticultural fleece. The tomatoes may then continue to ripen on the plants while being protected from a few degrees of frost. Take in the fruits if heavy frost is forecast.

The last of the crop

At the end of the season, hang trusses of tomatoes to ripen (opposite). There is a group of tomatoes known as long-keeper varieties, such as the Spanish 'De Colgar', which ripen slowly after harvesting if kept in a cool, frost-free place. They can take up to three months to reach maturity and should be ready for eating in winter, long after you have eaten your main-crop tomatoes.

There are occasional reports of tomatoes that remain in good condition for up to a year, but the mechanisms of fruit longevity are not fully understood at present.

RIPENING INDIVIDUAL TOMATOES – all you need is patience

Go bananas – put green tomatoes in a bowl next to a banana: it emits ethylene gas, which encourages the tomatoes to ripen.

RIPENING TRUSSES – so tempting they may be eaten off the line!

Cut trusses of green tomatoes and hang in an airy place such as a garage, potting shed, spare bedroom, or even the kitchen, to ripen.

Save your favourite seeds

If you have a favourite tomato that grows well in your garden, it is well worth saving the seeds for next year. Most tomatoes are self-pollinating, so the offspring will be identical to the parent plant.

However, if you or any of your neighbours grow several tomato plants within a bee's flight path, there is a risk that the plants will cross-pollinate. If so, the seedlings may vary from the parent. Currant tomatoes and many large beefsteak varieties have stigmas that protrude beyond the anthers, so are particularly vulnerable to cross-pollination. This may not worry you, but if you plan to donate seeds to a seed exchange, isolate the parent plant or bag individual trusses with a light muslin bag or an empty, dried tea bag, secured below the truss with a rubber band around the stem.

Tomato seeds are coated with a gel that protects them from attack by seed-borne diseases, like bacterial spot and early blight (pp114–19), but also inhibits germination. When preparing seeds for storage, you can remove the gel by fermentation.

STORING SEEDS – a cool, dark place keeps seeds viable for longer

Keep your seeds in labelled, envelopes or paper sachets (plastic bags encourage mould) with a sachet of silica gel to absorb moisture in an airtight container, in a cool, dark place, preferably in the salad drawer of the refrigerator.

DRYING – a simple method for a small amount of seeds

1 Collect seeds from ripe or even over-ripe fruits. Slice each tomato in half; scrape out the seeds with the point of a sharp knife.

2 To save only a few seeds, simply leave them to dry on a clean plate, but spread them out well so they do not go mouldy.

FERMENTING – good for large quantities of seeds

1 Scoop the pulpy seeds into a jar and leave in a warm place. After two days, a yeasty smell will indicate that the gel is fermenting.

2 Once the seeds are gel-free (a white fungal layer should form after 5–7 days), rinse the seeds. Spread out for a week to dry.

All for show

The record for the largest tomato was set in 1986 by Gordon Graham of Oklahoma, for a 'Delicious' weighing 3.51kg (7lb 12oz). Record-breaking or just for fun, showing tomatoes can be very satisfying.

Which class?

Usually, shows run by national organisations have traditional classes, but a local show may have a class for the ugliest or heaviest vegetable. Classes for 6–8 fruits require tomatoes of uniform size and colour, with intact, fresh, green calyces.

Grooming marvellous

Good soil preparation is important: work in plenty of organic matter to feed the plant. A sunny site is best for growth, but protect fruits from sunscald in very hot weather. Water regularly and apply a high-potash, or potassium (K), feed. Thin show tomatoes to five fruits per truss for the best shape and size. Support very large tomatoes; old stockings are stretchy and do not cut into the fruits. Protect ripening tomatoes from birds with netting or horticultural fleece.

It's all in the timing

Timing is vital in growing for exhibition. It is no good producing a perfect crop of evenly matched fruit if they do not ripen in the week of the show. Many variables affect maturation, chiefly sunlight and prevailing temperatures. The average number of days to maturity is often listed on seed packets, but these may vary greatly. The figure may be given as days from germination to first picking or as days from planting.

Exhibition growers often sow three or more batches of seeds at two-weekly intervals to ensure ripe tomatoes on the required day. Do not start your plants too early; even with grow-lights, it can be difficult to keep plants growing strongly. If you want giant fruits, expect to take 140–150 days from sowing seed to harvesting a ripe tomato.

A couple of days before the show, collect twice as many fruits as you need and place in a dark place to finish ripening. Select the best fruits on the day of the show. If fruit ripen too early, hold them back in the refrigerator at around 10°C (50°F) for up to three weeks; never freeze them, as they turn to mush on thawing. On the day, take care with labelling and presentation.

Varieties to show off

Good classic, round, F1 tomatoes include 'Cedrico', 'Cleopatra', and 'Vandos'. For giant fruits, try 'Slankards', 'Omar's Lebanese', or the hybrid 'Big Zac'.

Perfection Show fruits must be just ripe. >

Get grafting

Grafting is a way of joining two plants together so that they benefit from each other's strengths. It is fairly simple to carry out and is ideal if a particular tomato plant does not usually grow well for you.

Grafting allows a tomato variety to take on advantageous qualities of another tomato plant, usually a more robust species or variety. By grafting your chosen variety onto a rootstock, it is possible to promote growth, increase fruit quality and yield, and confer other benefits such as tolerance of low temperatures and resistance to soil-borne diseases. You may buy seed of tomato rootstocks, such as 'KNVF', 'Beaufort', and 'Brigeor'. Alternatively, use any tomato plant that you have found to be particularly resilient in your garden.

A graft should unite after 7–10 days: remove the cover and plant out or pot on the plant. Rub off any shoots emerging below the graft point (they will be more vigorous and overwhelm the grafted plant).

HOW TO GRAFT – combining the best qualities of two plants

1 Take the rootstock plant, and with secateurs or a clean, sharp knife, cut straight across to leave a short stem.

2 Make a vertical cut downwards about 2cm (½in) with a sharp knife into the centre of the rootstock stem.

3 Cut a healthy shoot from your chosen tomato plant. It should have a stem of a similar width to the rootstock.

4 Make two sloping cuts at the base of the stem, on opposite sides and about 2cm (½in) long, to create a tapered end.

5 Insert the tapered end of the shoot into the cut on the rootstock. Bind the graft with soft surgical or grafting tape or a grafting clip.

6 Cover with a plastic bottle or bag to keep humid and place in a cool, shaded position until the graft heals. Water regularly.

Create a variety of your very own

It is always fun to try to produce something new – so why not breed an heirloom tomato for your own family to treasure? Breeding tomatoes is not difficult and requires only a little patience.

You may identify traits that you would like, for example you might wish that 'Black Cherry' (p38) was available in a dwarf trailing form such as that of 'Tumbler' (p36). Crossing these two varieties and selecting the best offspring over a few generations may lead to a new plant with the desired characteristics. You may end up with many plants showing the growth habit of 'Black Cherry' and the fruits of 'Tumbler', but that is part of the fun.

The first step in breeding tomatoes is to cross-pollinate the two chosen parents (below). Collect seed from the female (seed-bearing) parent and raise new plants from them. Be ready to grow out many seedlings and keep crossing their offspring. If you do produce a plant worth keeping, isolate it and save seed for 3–5 years to ensure that it is stable. Then you can name it and, if you wish, offer it to commercial seed suppliers.

HOW TO CROSS-POLLINATE A TOMATO – create unique seeds

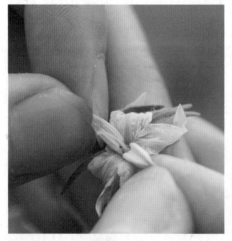

1 **Remove all anthers** (pp112–3) from a newly open flower with fingers or tweezers to create a seed-bearing parent.

2 **Rub the anthers** of the pollen parent over the stigma of the seed-bearing parent. Label with names of both parents.

Your perfect tomato plants Start a family
of plants to create your perfect crop. Watch
them grow, then have fun naming them.

In the kitchen

Store your tomatoes at room temperature to keep their wonderful flavour. Then, use the simple techniques at the beginning of this chapter to prepare, cook, and preserve them. The selection of recipes – from Cream of tomato soup to Tomato and Gruyère tart to Ice cream – will make the most of your home-grown crop. Try a dish perfect for zesty, green fruits or one of the recipes great for a glut. Enjoy!

Preparation

Preparing tomatoes starts with two things: a good wooden chopping board and a set of sharp knives for peeling, coring, slicing, and chopping. For large, meaty tomatoes, a potato peeler can also come in handy.

Paring knife This is ideal for coring and for skinning blanched tomatoes. You can also use it to help you scoop out any seeds.

Short blade and pointed tip

Small serrated knife You will use this time and time again. It cuts into tomato skins very easily without blunting.

Serrated blade

Chef's knife With its very sharp blade, this is great for slicing and chopping peeled tomatoes, as well dicing.

Broad fine-edged blade

SLICING – for all types of tomato

Sit the tomato on your chopping board and, using a serrated knife, cut into slices from the bottom to the top; discard the ends.

LONG-SLICING – for larger tomatoes

Using a chef's knife, cut into quarters and deseed (p139). Lie flat and slice thinly on the diagonal.

CHOPPING AND DICING – for all types of medium to large tomatoes

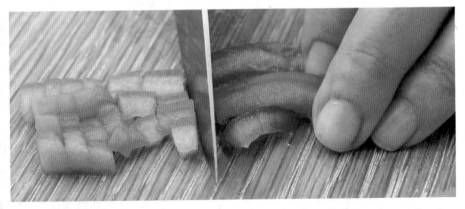

Once you have cut your tomatoes into long slices (above), gather all them under your hand, and turn so that they lie crossways to the knife. Roughly chop, or cut into dice as coarsely or finely as you need (the latter is known as a concassée.)

PEELING – for tomatoes to be eaten in juicy salads

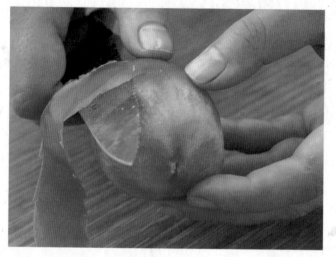

Using a serrated knife, make a small incision near the stem end. Peel carefully, rotating the tomato as you go, as though you were peeling an apple.

SKINNING BY CHARRING – very good for plum, globe, and cherry varieties

1 **Using tongs or a fork**, blister the tomato over a naked flame – take care. For a smoky taste, char on a barbecue or charcoal grill.

2 **When the tomato is black all over**, put in a plastic bag to sweat. Cool slightly, then skin using a blunt knife or kitchen paper.

SKINNING BY BLANCHING – ideal for plum and globe varieties

1 Cut a small cross on the bottom of each tomato. Blanch in very hot water for 15–20 seconds until the skins start to loosen.

2 Have a bowl of ice-cold water ready. Plunge the tomatoes into this to refresh. Remove with a slotted spoon, and allow to cool.

3 Once the tomatoes are cool enough to handle, remove the skins using a knife, as though you were peeling a banana.

CORING – useful for large, chunky tomatoes, which can have "woody" cores

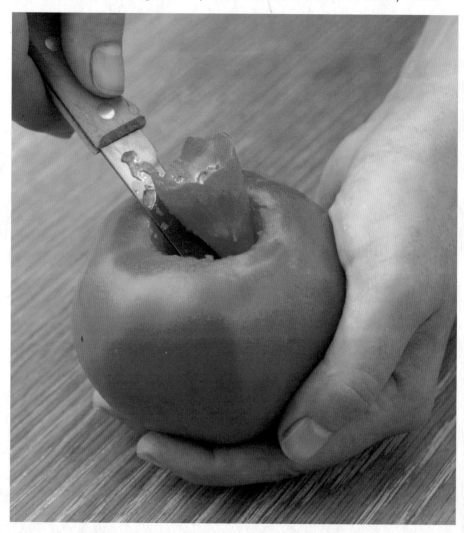

Use a paring knife to remove the core from tomatoes while they are still whole. Simply make a conical incision in the top of the tomato, slicing around in a circle, and pull out the core and discard.

DESEEDING – don't throw the seeds away, save them for making dressings

1 Use a paring knife to cut the tomatoes into quarters. Lie them flat on a chopping board and slice off the central core. The seeds and jelly-like juices will now easily slide out.

2 Using a teaspoon or your fingers, scoop out the seeds and jelly-like juice. If you do this over a bowl or jar, you can keep the juice for other uses, such as to add extra flavouring to a variety of dishes and to make dressings (p177). To discard the seeds, put in a sieve, and press to squeeze and separate.

Cooking

A portion of cooked tomatoes – simply fried, roasted, or grilled – makes an excellent accompaniment to many dishes. Use these three simple techniques to cook up a storm with your tomatoes.

The basic methods for cooking tomatoes can be applied and expanded on in so many ways, and tomatoes married with such a diverse combination of ingredients, that you could easily fill shelves and shelves with tomato recipe books. Hopefully, you will find some useful, easy, and delicious recipes to tempt you in the next section, but you can use these basic techniques as building blocks. Experiment with tomato varieties and with other ingredients. Trust your taste buds, and enjoy your time in the kitchen.

GRILLING AND CHAR-GRILLING – under the grill, in a pan, or on the barbecue

Char-grilling in a ridged cast-iron grill pan or on a griddle is a great method for plum, globe and even cherry tomatoes. Simply arrange the tomatoes in a very hot pan – there is no need to add oil. Leave to sizzle, turning from time to time, until they are charred but not burnt.

PAN-FRYING – halved, quartered, sliced, or chopped: the choice is yours

Pan-frying works well with globe, plum, and cherry tomatoes. Heat a little olive oil or butter, or a mixture of both, in a pan. Add the tomatoes, and sauté, stirring from time to time, until softened. Season, and serve. If you like, start by sautéeing some chopped onion first, then add a little garlic, and tip in your tomatoes. Sometimes simple is best.

ROASTING – choose your seasonings, drizzle with a little olive oil, and roast

Roasting is great for most tomato varieties. Line a roasting tin with greaseproof paper. Fill with halved, quartered, or even sliced tomatoes. Add peeled garlic cloves, lemon slices, and some thyme or oregano. Drizzle with a little olive oil, season, and roast in the oven at 180°C (350°C/Gas 4) for 25–30 minutes. For sweetly flavoured slow-roasted tomatoes, roast at 140°C (275°F/Gas 1) for at least 2 hours.

Preserving

Keeping otherwise perishable food for long periods of time is a very old technique, known as preserving. It is a great way of ensuring that you can enjoy the fruits in your garden all year round.

Sterilizing jars and bottles

When preserving, always sterilize jars first. Make sure that any lids will give a tight seal and use special preserving jars, such as Kilner jars, for whole tomatoes. Wash jars and lids in very hot, soapy water, rinse in boiling water, and put (minus any rubber seals) in a preheated oven at the lowest setting until dry. Carefully fill the hot jars to just under 2.5cm (1in) of the rim, and seal immediately.

When bottling whole tomatoes, you need to sterilize again as they cook. Stand the jars upright in a deep pan with an upturned heatproof plate or a wire rack set on the bottom. Cover with warm water until at least 2.5cm (1in) above the jars. Cover and boil gently for 40–50 minutes. Carefully remove the jars, and allow to cool overnight on a clean dry cloth. Store in a cool, dark place.

DRYING AND REHYDRATING – use any halved and deseeded tomatoes

1 To dry, roast skin-side down on a wire rack set over a baking tray at 75°C (150°F/Gas ¼) for 6 hours. Turn over, and roast for another 6 hours, with the door ajar or until very dry.

2 To rehydrate, soak dried tomatoes in a little hot water mixed with some olive oil for 5–10 minutes. Don't use too much liquid; if it is concentrated, it can be used later for cooking.

SMOKING – globes and plums

Put dried tomatoes on a wok grill over wood chips. Let smoke, covered, on medium heat, in a ventilated room, for about 15 minutes.

MAKING PRESERVES – spice on hand

Always sterilize jars (opposite) for chutney (p150), ketchup (p158), and jam before filling, sealing, and labelling.

BOTTLING – peeled globes and plums

Pack tomatoes to the neck of a sterilized jar; sprinkle with a little salt as you go. Add herbs and garlic. Preserve as given opposite.

FREEZING – suitable for sauce

Tomato sauce such as passata or marinara (p153) can be frozen in sealable plastic bags. Reheat until piping hot before serving.

Tomato borscht

Borscht comes in many variations, depending on its origins. In Russia and the Ukraine, it often includes tomatoes, as well as beetroot. This version may seem a bit unusual at first, but you will fall in love with its rich colour and fantastic taste – perfect for an elegant starter.

Serves 4
Prep time: 25 minutes
Cooking time: 25 minutes

2 tbsp olive oil
1 small onion, finely chopped
1 garlic clove, chopped
225g (8oz) raw beetroot, peeled and finely grated
1 tsp freshly ground toasted cumin seeds (see note)
½ tsp ground cinnamon
225g (8oz) ripe fresh tomatoes, skinned and roughly chopped
250ml (8fl oz) tomato juice
1 tbsp sun-dried tomatoes, very finely chopped
570ml (1 pint) vegetable stock
1 tbsp light soy sauce
sea salt and freshly ground black pepper

To serve
toasted cumin seeds
soured cream or crème fraîche

1 Heat the oil in a heavy pan over a low heat. Gently cook the onion and garlic for about 5 minutes, then add the beetroot. Sweat gently for a further 10 minutes, stirring from time to time, until softened but not browned.

2 Add the ground spices, fresh tomatoes, tomato juice, and sun-dried tomatoes, then pour in the stock. Bring to the boil. Reduce the heat slightly, cover, and simmer very gently for 15 minutes or until all the vegetables are soft. Remove from the heat. Blend or process until velvety smooth. Check the seasoning, adding the soy sauce, salt, and pepper to taste.

3 Serve chilled, at room temperature, or slightly warm. If you do reheat the soup, do so gently over a low heat. To serve, spoon into serving bowls, and garnish with toasted cumin seeds and a spoonful of soured cream or crème fraîche.

Note For the best, most flavourful results, toast whole cumin seeds in a dry frying pan over a medium heat for a few minutes until golden and aromatic. Grind to a powder using a mortar and pestle. Use straight away.

Eva's Purple Ball (p31)

Works well with
Standard globe (pp22–35)
Plum (pp64–73)

Cream of tomato soup

In this extra-special version of the old standard, the tomato is the star. Here we have tomatoes in all variations: fresh, sun-dried, and roasted. Inspired by tomato-lovers, this velvety smooth soup takes the humble tomato to a dimension where it rightfully belongs.

Serves 4–6
Prep time: 30 minutes
Cooking time: 40 minutes

50g (1³⁄₄oz) butter
1 tbsp olive oil
2 onions, finely chopped
2 celery sticks, finely chopped
2 carrots, finely diced
2 garlic cloves, minced
12 plum tomatoes, about
   1kg (2¹⁄₄lb), quartered,
   roasted (p156), and
   roughly chopped, plus 8
   extra 600–720g (1¹⁄₄–1¹⁄₂lb),
   skinned and finely chopped
6 sun-dried tomatoes,
   finely chopped
1 litre (1³⁄₄ pints) hot
   vegetable stock
2–3 tbsp double cream
sea salt and freshly ground
   black pepper

1 Heat the butter and olive oil in a heavy saucepan over a medium heat. Add the onions, and sauté for 8–10 minutes, stirring frequently, until very soft but not coloured. Next, add the celery and carrots, and continue cooking gently without burning for another 10 minutes. Lastly, add the garlic, and sauté for another 2 minutes, all of this while stirring from time to time.

2 Mix together the roasted plum tomatoes, sun-dried tomatoes, and fresh tomatoes. Tip into the pan with any juices, and cook, stirring, for 5 minutes so that the flavours combine; if the sauce looks too thick or starts catching on the bottom of the pan, add a little of the hot vegetable stock. Pour in the remaining vegetable stock, and simmer the soup for 15–20 minutes.

3 Blend the soup to a smooth purée using a food processor or hand-held blender. Pass through a sieve or mouli into a clean pan. Add the double cream a teaspoon at a time until you are happy with the taste and texture. Season with salt and pepper, reheat very gently if needed, and serve.

Variation For a peasant-style soup, don't pass the soup through a sieve or mouli. Simply stir in a little double cream, and season to taste.

Works well with
Plum (pp64–73)
Beefsteak (pp52–63)
Standard globe (pp22–35)

Salsa romesco

Romesco, Catalan in origin, is a pounded sauce that stars two key ingredients: tomatoes and romesco peppers. In Catalunya, festivals are held to celebrate this dish, and families hold their recipes as closely guarded secrets. Be sure to use the choicest, ripest tomatoes.

Makes 1 large bowl
Prep time: 25 minutes
Cooking time: 30 minutes

1 dried nyora (ñora) pepper
  or, for variety, any other hot
  dried pepper such as cascabel,
  guajillo, or pimento
20 blanched almonds
2 x 2.5cm (1in) pieces of day-
  old baguette
1 large beefsteak tomato, about
  280g (10oz), skinned,
  quartered, and deseeded
3 garlic cloves, unpeeled
60ml (2fl oz) red wine
4 tsp red wine vinegar
about 100ml (3½fl oz) good-
  quality olive oil
sea salt and freshly ground
  black pepper

1 Preheat the oven to 180°C (350°F/Gas 4).

2 Deseed the nyora or other dried pepper, and use your hands to tear the flesh into small pieces. Put in a bowl with 3 tablespoons boiling water, and leave to infuse for about 15 minutes. Drain off the soaking water, reserving the chilli.

3 Meanwhile, put the tomatoes, bread, garlic, and almonds in a roasting tin, sprinkle with a little olive oil, and roast in the oven. Keep an eye on things because the various ingredients take different amounts of time to roast. Allow 5–8 minutes for the almonds, 10 minutes or so for the bread, and about 15 minutes for the tomatoes and garlic.

4 Using a mortar and pestle, pound together the peppers and roasted tomatoes, bread, garlic, and almonds until you have a thick paste; alternatively, whiz in a food processor. Add the red wine and the vinegar, then gradually add about 100ml (3½fl oz) olive oil, combining as you go, until you have a rich, dark sauce. Season with salt and pepper. Enjoy with grilled meats and seafood, or add to soups and other dishes for an extra kick, or to mayonnaise for added spice.

Note To make your *salsa romesco* even richer, add 2 or 3 roughly chopped sun-dried tomatoes. If using sun-dried tomatoes stored in olive oil, simply drain first, and pound with the other ingredients. Sun-dried tomatoes not stored in oil should be soaked with the nyora peppers before using.

Works well with
Beefsteak (pp52–63)
Plum (pp64–73)

Pasta with fresh tomato sauce

A herby "pesto" or "picada", made with sun-dried tomatoes, transforms this sauce into something really memorable. Try adding 2 finely chopped anchovies (preferably those preserved in oil) and some very thinly sliced fennel bulb, to enliven the sauce even more.

Serves 4–6
Prep time: 15–20 minutes
Cooking time: 10 minutes

2 tbsp good-quality extra virgin olive oil

2 roasted garlic cloves (see note)

2 tbsp roughly chopped flat-leaf parsley

2 tsp dried oregano or 1 tsp fresh

½ tsp *pimentón picante* (Spanish hot smoked paprika)

3 sun-dried tomatoes in olive oil, drained (reserve the oil)

4 large ripe tomatoes, about 400g (14oz), chopped

2 x 400g (14oz) cans good-quality peeled plum tomatoes, drained and roughly chopped

generous pinch of light soft brown sugar

2 tbsp sherry vinegar

500g (1lb 2oz) dried pasta such as spaghetti, fusilli, fregula, orecchiette, or tortiglioni

100g (3½oz) fresh rocket leaves

sea salt and freshly ground black pepper

freshly grated or shaved Parmesan cheese, to serve

1 Using a mortar and pestle, pound the olive oil, roasted garlic, parsley, oregano, paprika, and sun-dried tomatoes to a paste that resembles pesto. Alternatively, whiz briefly in a small blender until a rough pesto forms. If you need to, add a little of the reserved oil from the sun-dried tomatoes.

2 Mix together the fresh and canned tomatoes, sugar, and vinegar. Season with salt and pepper, and stir through.

3 Cook the pasta in plenty of boiling salted water, according to the packet instructions, until *al dente*. Drain, reserving a little of the cooking water to keep the pasta moist.

4 Meanwhile, transfer the tomato mixture to a hot frying pan over a medium heat. Add the pesto ingredients. Stir for a minute or so to *just* warm through, then remove from the heat. Quickly toss through the freshly cooked pasta, and stir in the rocket. Serve immediately, sprinkled with Parmesan.

Note Roasting garlic is very simple. Just put a couple of fat unpeeled cloves on a hot griddle or ridged cast-iron grill pan. Char-grill for a few minutes on each side until soft inside. Allow to cool, peel, and use the roasted pulp as in the recipe.

Hillbilly Potato Leaf (p53)

Works well with
Beefsteak (pp52–63)
Standard globe (pp22–35)

Chutneys

The first of this spicy duo is a fresh Hyderabadi chutney that is at its best when eaten straight away. *Pimentón* and roasted red peppers add smoky piquancy to the second, which only improves over time.

Makes about 1 kg (2¼lb)
Prep time: 15 minutes
Cooking time: 20 minutes

800g (1¾lb) cherry tomatoes
1 tsp chilli powder
1 tsp ground turmeric
2 garlic cloves, crushed to
  a paste with a little salt
100ml (3½fl oz) vegetable oil,
  plus 3 tbsp extra
2 tsp cumin seeds
3 dried red chillies
10 fresh or frozen curry leaves

Savi's fresh tomato chutney (pictured)

Put the tomatoes, chilli powder, turmeric, and garlic in a large stainless-steel saucepan over a medium heat. Bring to a simmer, and bubble until the liquid evaporates – about 10 minutes. Add the 100ml (3½fl oz) oil to the pan, and continue cooking until the oil rises to the top. Next, temper the spices. Heat the extra oil in a small pan or wok over a medium-high heat. When the oil is hot, add the cumin seeds, then the chillies, and finally the curry leaves. Once the chillies have turned dark red, empty the contents of the pan into the tomato chutney. Mix well. Be careful, as the spices can burn very quickly. Serve as part of an Indian meal or just as an excuse to spice up most things – it is very moreish indeed. It will keep in the refrigerator for up to a week.

Makes about 1.5 kg (3lb 3oz)
Prep time: 40 minutes
Cooking time: 1–1½ hours

1kg (2¼lb) ripe tomatoes,
  skinned, cored, and roughly
  chopped
250g (9oz) onions, roughly
  chopped
2 fresh red peppers, roasted,
  skinned, and roughly chopped
100g (3½oz) demerara sugar
2 tsp salt
1 tsp pimentón picante
  (Spanish hot smoked paprika)
300ml (10fl oz) red wine vinegar

Tomato chutney with pimentón and red peppers

Put all the ingredients, except for the vinegar, in a large stainless-steel pan. Cook over a medium heat until the sugar starts to dissolve and coats the rest of the ingredients. Add the vinegar, and bring to the boil. Reduce the heat to a gentle simmer, and cook for 1–1½ hours, stirring from time to time, until everything is very soft and you have a thick consistency. Pot into hot, sterilized jars with tight-fitting lids, seal straight away, and label (pp142–3). This chutney is perfect with grilled meats. Its flavours improve if allowed to mature before opening, and it will keep for up to 9 months if stored in a cool, dark place.

Gazpacho

Ideal for when you have a glut of tomatoes on your hands, the truly great thing about gazpacho is how easy it is to make. The only thing you need for an irresistible result is good-quality ingredients – they really do make all the difference to this refreshing chilled soup.

Serves 4
Prep time: about 20 minutes
Cooking time: 5–7 minutes
Chilling time: 2 hours

4 large ripe beefsteak tomatoes, about 1–1.1kg (2¼–2½lb), skinned, quartered, and deseeded

3 tbsp extra virgin olive oil

1 tbsp sherry vinegar, cider vinegar, or apple vinegar

1 tbsp crumbled chicken or vegetable stock cubes

2 green peppers, deseeded and finely diced

1 large cucumber or 2 Lebanese cucumbers, peeled, deseeded, and finely diced

sea salt and freshly ground black pepper

To serve

garlic croutons (see cook's tip)

a couple of hard-boiled eggs, diced (optional)

1 Put the tomatoes in a blender with a little water, and blend to a very smooth pulp. Transfer the pulpy liquid to a heavy saucepan, and heat gently for 2–3 minutes.

2 Add the olive oil, vinegar, crumbled stock, peppers, and diced cucumber. Heat over a medium heat for 3–4 minutes until just below boiling point, stirring from time to time. Remove from the heat, and allow to cool.

3 Once the soup has cooled to room temperature, season very well with salt and pepper; don't stint on the seasoning. Pour the soup into a large bowl, and chill in the refrigerator for a few hours until cold. Taste and season again if needed.

4 Serve the gazpacho chilled, garnished with garlic croutons and diced hard-boiled egg (if using).

Note It is important that your gazpacho is well seasoned – almost too generously. As it is going to be eaten cold, it needs to be very tasty. Remember, the flavours in cold food do not come through as strongly as they do in food that is served hot or at room temperature.

Cook's tip To make garlic croutons, rub some day-old bread with the cut side of a garlic clove. Cut into small cubes, toss in a little olive oil, and sprinkle with sea salt. Spread out on a baking tray, and roast in a preheated 180°C (350°F/Gas 4) oven for 10 minutes or until crisp and golden.

Works well with
Beefsteak (pp52–63)
Plum (pp64–73)

Simple marinara-style sauce for pizza

The name *marinara* for the classic Italian sauce comes not from its contents, but from the fact that it was a favoured choice, atop a round of flat bread, for hungry fishermen in Naples returning to port after a long night at sea. All those healthy appetites can't be wrong ...

Makes enough for
2 medium pizzas
Prep time: 10 minutes
Cooking time: 25 minutes

5 ripe tomatoes, about
  450g (1lb), skinned
  and chopped

2 tbsp extra virgin olive oil

2 garlic cloves, crushed

½ tsp dried oregano

1 tbsp capers in brine,
  rinsed, gently squeezed dry,
  and finely chopped

2 tbsp roughly chopped
  flat-leaf parsley

1 Put all the ingredients except for the parsley in a saucepan, and simmer over a low heat for about 5 minutes until the tomatoes become watery. Cook, covered, for about 15 minutes until the tomatoes and garlic are soft. Add the parsley, and cook for a further 5 minutes

2 Remove from the heat, and allow to cool. Spread onto your pizza as a base for other toppings of your choice. Or, if you want to serve your pizza in the classic marinara style, simply spread the pizza dough with the topping, scatter a little slivered garlic and a sprinkling of oregano over the top if you wish, drizzle over some extra virgin olive oil, and bake. And there you have it – a version of the original pizza topping so beloved of Neapolitans.

Note Why not make a batch of this sauce and freeze it to use later? After cooking the tomatoes and garlic in step 1, pour the sauce into a sealable plastic bag and freeze (p143).

Works well with
Plum (pp64–73)
Beefsteak (pp52–63)
Standard globe (pp22–35)

Salsas

Salsa de molcajete has a wonderful smoky flavour from the char-grilling, while *Pico de gallo*, which literally means "rooster's beak", is the mother of all Mexican salsas with its fresh simplicity.

Makes about 500g (1lb 2oz)
Prep time: 10 minutes
Cooking time: 10 minutes

6 plum tomatoes or
  4 beefsteak tomatoes
2 fresh green bird's-eye
  chillies, stalks removed and
  roughly chopped, or
  to taste
½ small red onion, finely
  chopped
3 tbsp finely chopped coriander
  (use leaves and stalks)
juice of 1 lime
sea salt

Salsa de Molcajete (pictured)

Put the tomatoes and chillies on a very hot griddle or under a very hot grill, and blacken on all sides. Allow to cool for about 5 minutes. On a wooden chopping board, cut the tomatoes into quarters, then tip into a food processor with their seeds and any juice. Add the chillies, and process for about 30 seconds until you have a coarse-textured purée. Alternatively, pound the tomatoes and chillies using a large mortar and pestle. Transfer to a serving bowl, and add the onion, coriander, and lime juice. Season generously with salt. Taste, and adjust as needed – remember, seasoning is key, so don't stint. Leave to stand for 1 hour to allow the flavours to develop. This salsa also keeps very well for a couple of days if stored, covered, in the refrigerator. Serve at room temperature with tacos, fajitas, or char-grilled meats.

Makes about 350g (12oz)
Prep time: 10 minutes

6 plum tomatoes, deseeded and
  finely chopped
1 white onion, finely chopped
1 ripe avocado (optional)
3 tbsp finely chopped coriander
  (use both leaves and stalks)
1–2 fresh bird's-eye, hot green,
  or serrano chillies, deseeded
  and finely chopped
juice of 1 lime
sea salt

Pico de gallo

Put the tomatoes and onion in a bowl. Halve and stone the avocado (if using), and cut into dice. Add to the bowl with the coriander, and stir through gently. Next, add the chilli a little at a time, tasting as you go, and stop when you reach the desired level of heat. Season with a little salt, and squeeze over some fresh lime juice. Serve immediately with freshly made tortillas, in tacos or fajitas, or alongside minute steaks marinated with lime juice and a little olive oil.

Variation Stir some sweetcorn kernels or cooked black beans into the salsa just before serving.

Roasted tomato and garlic soup

Simple to make ... deliciously satisfying results. If you like, you can roast the vegetables in advance, and keep them in the refrigerator for 2–3 days until needed; the really good news is that, once this is done, the soup takes only 5 minutes to prepare. Home-made food in a jiffy.

Serves 4
Prep time: 15 minutes
Cooking time: 30 minutes

8 plum tomatoes, about 675g (1½lb), quartered

1 red onion, cut in 8 wedges

2 garlic cloves

3 tbsp olive oil

1 litre (1¾ pints) hot vegetable stock

3 tbsp sun-dried tomato paste

sea salt and freshly ground black pepper

1 To roast the vegetables, heat the oven to 180°C (350°F/ Gas 4). Put the tomatoes, onion, and garlic on baking trays covered with greaseproof paper. Drizzle with a little bit of olive oil, and season well with salt and pepper. Roast until all the vegetables are soft, caramelized, and slightly browned – allow 10–15 minutes for the garlic, 15–20 minutes for the onion, and 25 minutes for the tomatoes. Squeeze the garlic from the skin once it has cooled slightly.

2 Using a hand-held blender, whiz together the vegetable stock, sun-dried tomato paste, and roasted vegetables until puréed, but still with some chunky bits. Adjust the seasoning as needed, and reheat gently. Serve hot.

Variations Try adding a dollop of mascarpone cheese, croutons, or roasted cumin seeds to each serving. Or vary the soup recipe itself by adding some chopped fresh herbs such as thyme, basil, or rosemary.

Works well with
Plum (pp64–73)
Cherry (pp36–51)

Sofrito

Sofrito, sofrigit, soffritto, refogado – this method for a base sauce pops up in Spanish, Latin American, Italian, Portuguese, and Sephardic cuisine. There are countless variations, and each cuisine has its particular stamp – onion, garlic, and tomatoes in Spain, for instance.

Makes about 300g (10oz)
Prep time: 5 minutes
Cooking time: 25 minutes

Basic sofrito

1 onion, finely chopped or finely sliced

1–2 tbsp good-quality olive oil

1–2 garlic cloves, finely chopped or pounded to a paste using a mortar and pestle

4 ripe plum tomatoes, about 300–360g (10–12oz), chopped (peeled and deseeded, or not, according to your taste)

1 Heat the oil in a medium frying pan. Add the onion, and gently sauté for 5–10 minutes, stirring from time to time with a wooden spoon, until golden brown. Add the garlic, and sauté for another minute, stirring continuously.

2 Tip in the tomatoes, and cook over a medium heat, stirring from time to time. As the tomatoes cook down, the sauce becomes syrupy and rich. Depending on the type of dish you are making, you may need to cook for 15 minutes or longer.

3 At this point, the basic sofrito is ready. You can now add a variety of herbs, spices, meat, fish, or wine, then continue with your dish. Below are some suggestions.

Variations

• Deseed a dried ancho or chipotle chilli, and soak in a little hot water. Blend the chilli and its soaking liquid to a paste. Add to the sofrito, to make a Mexican salsa for meatballs or use for dishes such as chilli con carne.

• Add a glass of good red wine and add some fresh herbs such as rosemary or oregano, to use with pork or lamb.

• Soak a few saffron threads in 100ml (3½fl oz) water. Add to the sofrito, and reduce slightly. Use with pasta shapes such as orzo, fregola, and malloredus (gnochetti sardi).

• Add ½ teaspoon pimentón (Spanish smoked paprika) and a little freshly squeezed orange juice – great with white fish.

Works well with
Plum (pp64–73)
Standard globe (pp22–35)

Ketchup ... of course

A book on tomatoes would not be complete without ketchup ... so here is a recipe for classic tomato ketchup, in trademark red, with a green tomato variation, great for using up lots of unripe tomatoes.

Makes about 2kg (4½lb)
Prep time: 40 minutes
Cooking time: 2 hours

2kg (4½lb) ripe plum tomatoes, skinned and roughly chopped

500g (1lb 2oz) Granny Smith or other cooking apples, peeled, cored, and chopped

3 onions, chopped

150g (5oz) white granulated sugar

70g (2½oz) soft dark brown sugar

500ml (16fl oz) cider vinegar or apple vinegar

1 tbsp salt

½ tsp chilli powder

½ tsp ground cinnamon

½ tsp freshly ground black pepper

8 allspice berries

7 whole cloves

Red tomato ketchup (pictured)

Put all the ingredients in a large stainless-steel pan. Bring to the boil, then reduce the heat slightly. Simmer, uncovered, for 2 hours, stirring from time to time. Remove from the heat, and allow to cool slightly. Using a hand-held blender, purée the mixture until velvety smooth. If the sauce is still watery, simmer for another 20–30 minutes until thick and a rich reddish brown, stirring all the time. Pass the ketchup through a sieve, and pot into hot sterilized jars or bottles. Seal tightly with vinegar-proof lids straight away, then label and store in a cool, dark place. It keeps for up to 9 months.

Green tomato ketchup

Use either unripe tomatoes or green varieties such as 'Green Zebra' (p26) for this variation, and perhaps even a few tomatillos (papery husks removed). You will end up with about 1kg (2¼lb) ketchup. Put 1.5kg (3lb 3oz) cored and quartered unripe or green tomatoes, 500g (1lb 2oz) Bramley or other cooking apples, peeled, seeded, and cut into small dice, and 1 large chopped onion in a large stainless-steel pan. Add 1 cinnamon stick, 6 allspice berries, 4 cloves, ½ tsp cayenne pepper, 1 tsp black peppercorns, 1 bay leaf, 1 tbsp salt, and 300ml (10fl oz) cider vinegar or apple vinegar. Bring to the boil, and cook for 20 minutes or so until the fruit and vegetables soften. Add 200g (7oz) demerara sugar, dissolve completely, then bring to a gentle simmer. Cook for 1 hour, uncovered, until very soft and thick, stirring from time to time. Blend to a smooth purée, then pass through a sieve back into the pan. Bubble for another 20–30 minutes until the ketchup is thick. Pot, seal, and store as above. It keeps for up to 6 months.

Tomato and Gruyère tart

The flavours of tomatoes and Gruyère combine very well. Gruyère also pairs beautifully with white wine, so the white wine in the pastry makes this tart a good combination altogether.

Serves 8
**Prep time: 30 minutes plus
30 minutes for chilling**
Cooking time: 45 minutes

For the pastry

250g (9oz) plain flour

pinch of salt

100g (3½oz) cold butter,
cut into cubes

25g (scant 1oz) cold or
frozen lard or Trex (vegetable
cooking fat)

1 egg

about 4 tbsp white wine

For the filling

200ml (7fl oz) double cream

200g (7oz) Gruyère cheese,
grated

50g (1¾oz) plain oatcakes,
crushed into crumbs

about 800g (1¾lb) ripe
tomatoes, halved and
deseeded

½ tsp dried thyme or oregano

freshly ground black pepper

1 To make the pastry, whiz the flour, butter, lard, and salt in a food processor for about 20 seconds until resembling breadcrumbs, then add the egg. Keep whizzing, adding the wine a little at a time, just until the dough comes together. With your hands, quickly work into a disk about 2.5cm (1in) thick. Wrap in cling film, and chill for about 30 minutes.

2 Preheat the oven to 180°C (350 F°/Gas 4).

3 Roll out the pastry on a floured work surface until about 5mm (¼in) thick. Use to line a 25cm (10in) loose-bottomed flan or tart tin. Prick the base with a fork, and line the tin with baking parchment. Fill with ceramic beans or uncooked pulses, and bake blind for 10 minutes. Remove the paper and beans, and bake the pastry for a further 5 minutes.

4 Spread the oatcake crumbs over the bottom of the pastry case, and top with the tomatoes. Sprinkle over the thyme or oregano, and season with salt and black pepper. Scatter with the cheese, then pour over the cream.

5 Bake in the middle of the oven for 30 minutes or until the tart looks brown all over. Leave to cool slightly before removing from the tin. Serve hot or cold with a salad.

Loveheart (p41)

Works well with
Cherry (pp36–51)
Standard globe (pp22–35)

Basil and Parmesan stuffed tomatoes

It simply had to be: a recipe for this culinary stalwart of tomato cuisine. These are great served on their own or even grilled in their foil parcels as part of a barbecue. Remember, always use fresh herbs.

Makes 1 large stuffed tomato*
Prep time: 25 minutes
Baking time: 30 minutes

1 large beefsteak tomato, about 250–280g (8–10oz)
1 large egg, beaten
1 tbsp fresh white breadcrumbs
1 tbsp olive oil
½ tsp sugar

Basil and Parmesan filling

2 tbsp shredded basil leaves
1 tbsp finely chopped flat-leaf parsley
1 tbsp finely chopped rinsed and drained capers
1 tbsp freshly grated Parmesan cheese
½ tsp freshly grated lemon zest
salt and freshly ground black pepper
pinch of sugar
salt and freshly ground black pepper

* It is easy to make this recipe for more people. Simply adjust the amounts according to how many tomatoes you are using and how big they are.

1 Preheat the oven to 180°C (350°F/Gas 4). Carefully slice the top off the tomato so that it has a lid. Core and scoop out the insides, removing the seeds and jelly-like juice (p138). Put the tomato shell and its lid cut-side down in a colander.

2 To make the filling, mix together the basil, parsley, capers, Parmesan, and lemon zest. Season generously with salt and pepper, taste, and adjust the seasoning if needed. Combine with the beaten egg and breadcrumbs. Fill the tomato shell until it is well stuffed, cover with its lid, and place right-side up in a roasting dish lined with a piece of foil.

3 In a small bowl, mix together the oil, sugar, and 2 tbsp water. Pour over the tomato, wrap over the foil to make a parcel, and bake in the middle of the oven for 30 minutes or until soft and gooey. Serve on its own, accompanied by a peppery salad of rocket and watercress, or a side dish.

Variations Use 1 tbsp roughly chopped tarragon, 1 tbsp finely chopped flat-leaf parsley, 1 tbsp finely chopped rinsed and drained capers, 2 tsp freshly grated lemon zest, and 1 tbsp single cream for the filling. Season, and make as above. Or try mixing the scooped-out flesh, seeds, and jelly-like juice with 1 tbsp chopped dill, 1 tbsp shredded mint leaves, 1 tbsp finely chopped onion or shallot, 1 tbsp chopped toasted pine nuts, 1 tsp freshly squeezed lemon juice, 1 tbsp olive oil, and a pinch of sugar. Mix together, season, and make as above. For an extra, meaty dimension to the latter filling, add some sautéed lamb mince.

Work well with
Beefsteak (pp52–63)
Standard globe (pp22–35)

Ragù bolognese

This unashamedly tomato-heavy interpretation of the classic *ragù bolognese* comes from Marine Ices, a London culinary institution. Long, slow simmering brings out the rich flavour of the tomatoes. Thank you, Marine Ices.

Serves 8
Prep time: 45 minutes
Cooking time: at least 2 hours

1 tbsp olive oil
1 onion, finely chopped
2 celery sticks, finely chopped
2 carrots, finely diced
1kg (2¼lb) 10% fat beef mince
175ml (4½fl oz) red wine
3kg (6½lb) ripe plum tomatoes, skinned and chopped
200g (7oz) double-concentrate tomato purée
1 bay leaf
freshly ground nutmeg, to taste
1 tbsp butter
salt and freshly ground black pepper

1 Heat the oil in a very large saucepan, and add the onions, celery, and carrots. Sweat gently for 5–10 minutes until soft but not browned. This initial technique is an example of an Italian *soffritto* (p157).

2 Add the mince, and cook gently, until the meat is well browned, stirring constantly and breaking up the meat with a wooden spoon.

3 Pour in the red wine, then add the tomatoes, tomato purée, and bay leaf. Stir through, then bring to just below the boil. Reduce the heat, and gently simmer the ragù for at least 2 hours, stirring from time to time, until well reduced to a nice thick sauce.

4 Season well with a good pinch of nutmeg, salt, and pepper. Taste, adjusting the seasoning if needed, then stir in the butter. Serve hot with pasta or use as a base for lasagne.

Note The time spent skinning the tomatoes for this recipe may seem a little daunting – but persevere. Your efforts will be well rewarded in the delicious end result.

Works well with
Plum (pp64–73)

Mexican-style tomato summer pudding

Make this sort of treasure chest brimful of tasty tomatoes – green ones, yellow ones, red ones, quartered plum tomatoes, long slices of beefsteaks, tomato wedges, even whole pricked cherry tomatoes.

Serves 4
Prep time: 30 minutes
Soaking time: overnight

1 small loaf sliced white bread, crusts removed

8–9 very ripe tomatoes, about 675g (1½lb), cored and deseeded (reserve the seeds and jelly-like core)

2 roasted tomatoes, cut into quarters (p141)

3 sun-dried tomatoes in olive oil, chopped

300ml (10fl oz) tomato juice

5 tbsp olive oil

grated zest (optional) and juice of ½ lime

good squirt of Maggi Liquid Seasoning

good squirt of light soy sauce

good squirt of Worcestershire sauce

2 tbsp home-made or good-quality ketchup (p158)

1 fresh red bird's-eye chilli, finely chopped

4 tbsp finely chopped coriander leaves

1 tbsp finely chopped shallot

200g (7oz) peeled and deveined cooked prawns

sea salt and freshly ground black pepper

freshly sliced ripe avocado

1 Set 2 whole bread slices aside, and cut the rest in half. Line a 1.5-litre (2¾-pint) glass or ceramic bowl with the bread halves, overlapping each slice a little bit.

2 Cut the fresh tomatoes into random shapes – peel some, leave others unpeeled, and keep small ones whole. Set aside with the roasted and sun-dried tomatoes.

3 Blend or process the reserved seeds and "jelly" to a purée, and put in a jug. Add the remaining ingredients, except for the prawns and avocado. Do this little by little, tasting as you go. The flavour should be strong but not overpowering. Season with salt and pepper.

4 Mix the prepared tomatoes and prawns together, and use to fill the bread bowl – they should come three-quarters of the way up the sides. Pour over half of the spicy juice mix until the tomatoes are covered. Use the reserved 2 whole slices of bread to make a "lid" on top. Carefully pour over the remaining spicy juice mix, place a saucer directly on top of the bread, weigh down heavily, and chill overnight.

5 To serve, turn the bowl upside down onto a large plate. Carefully lift off, so that the pudding breaks open and people can help themselves. Serve with the avocado slices.

Yellow Butterfly (p68)

Works well with
All tomatoes (pp22–77)

Baked beans

If you're looking for inspiration, try this dish. It takes a little planning, as you need to soak the beans overnight, but after that everything is simply put together and baked. Once you have tasted these, you will realize exactly how good home-made baked beans really are.

Serves 4
Soaking time: overnight
Prep time: 20 minutes
Cooking time: 2–3 hours

250g (9oz) dried haricot beans
1 tbsp vegetable oil
1 onion, finely chopped
2 garlic cloves, finely minced
1 tbsp black treacle
1 tbsp maple syrup
½ tsp ground cumin
¼ tsp ground cinnamon
¼ tsp cayenne powder
5 large tomatoes, about
   450g (1lb), skinned
   and chopped
1 tbsp very finely chopped
   sun-dried tomatoes
1 tbsp tomato purée
1 tsp English mustard
sea salt and freshly ground
   black pepper

1 Put the haricot beans in a large bowl, and cover with plenty of cold water – enough to cover by at least 4cm (1¾in). Leave to soak overnight.

2 Drain and rinse the soaked beans. Put in a large heavy pan, cover with plenty of fresh cold water, and bring to the boil. Boil rapidly for 10 minutes, reduce the heat slightly, and simmer for 1 hour; drain.

3 Preheat the oven to 140°C (275°F/Gas 1). Put the beans in a flameproof casserole with the remaining ingredients, except for the salt (this is important – adding salt at this point toughens the beans). Season with a good grinding of pepper, and add 500ml (16fl oz) cold water.

4 Gently bring to the boil, cover, and transfer the casserole to the oven. Bake for about 4 hours, stirring from time to time. Keep an eye on the beans as they cook; if the mixture is too thick or starting to dry out, top up with a little water as needed. Once the beans are creamy and soft, add enough salt to season well. Bake for another 5 minutes, then serve.

Variation To cook the beans without baking, follow steps 1 and 2, then put the beans in a clean pan with the rest of the ingredients, except the salt. Season with pepper, and bring to the boil. Reduce the heat slightly, cover, and simmer very gently, stirring from time to time, for 1–2 hours until the beans are tender. Season well with salt, and serve.

Works well with
Plum (pp64–73)
Standard globe (pp22–35)

Slow-cooked ox cheeks with braised vegetables

Arte Culinaria is a fantastic cookery school in Italy's Veneto region, and its hostess, Antonella Tagliapietra, is the source of this wonderful dish. A classic example of slow cooking, it uses a cut of meat that is often overlooked – truly mouthwatering and meltingly tender.

Serves 4
Prep time: 20 minutes
Cooking time: 3 hours

3 tbsp olive oil

2 onions, chopped

700g (1½lb) ox cheeks, cut into chunks and trimmed of any fat

1 bay leaf

sprig of fresh rosemary

150ml (5fl oz) dry white wine

2 large carrots, thickly sliced

1 celeriac, peeled, quartered, and sliced

6–7 very ripe tomatoes, about 500g (1lb 2oz), skinned and chopped

1 tsp salt

1 Heat the oil in a large heavy pan such as an enamelled cast-iron saucepan or casserole over a medium-high heat. Sauté the onions until starting to turn golden brown.

2 Add the ox cheeks, bay leaf, and rosemary to the onions, and keep stirring until everything starts to brown. When the meat begins to stick to the bottom of the pan, splash in the wine, and stir to deglaze.

3 Next, add the carrots, celeriac, tomatoes, salt, and 250ml (8fl oz) water. Reduce the heat to low, stir through, and leave the meat and vegetables to simmer very gently, uncovered, for 2½ hours until tender. Stir from time to time, so that they do not catch on the bottom, topping up with a little extra wine or water if the mixture becomes too dry.

4 Serve hot with polenta.

Variation Beef shoulder and neck are also good choices for this recipe. You could also try it with cuts of lamb that benefit from long, slow cooking.

Works well with
Plum (pp64–73)
Standard globe (pp22–35)

Tomato panade

This is a baked combination of the traditional French soup *panade*, usually eaten as a thick bready broth with caramelized onions, and the Italian *pappa al pomodoro*, another bready broth, but with tomatoes.

Serves 6 as a main course and 8 as a side dish
Prep time: 40 minutes
Cooking time: 40 minutes

3 large onions, finely sliced

about 2 tbsp olive oil, plus extra for drizzling

8 garlic cloves, roughly chopped and crushed to a paste with a little salt

½ fresh red chilli, deseeded and very finely chopped

4 ripe plum tomatoes, about 300–360g (10–12oz), roughly chopped, or 1 x 400g (14oz) can whole peeled plum tomatoes, roughly chopped

400g (14oz) baguette, cut into slices 5mm (¼in) thick

handful of fresh basil leaves, roughly torn

6 large, ripe round tomatoes, or 2–3 beefsteak tomatoes, about 550g (1¼lb), cut into 5mm (¼in) slices

50g (1¾oz) Parmesan cheese, freshly grated

125g (4½oz) mozzarella cheese, sliced

300ml (10fl oz) hot chicken stock

sea salt and freshly ground black pepper

1 Preheat the oven to 180°C (350°F/Gas 4).

2 Heat about 2 tbsp olive oil in a large frying pan over a medium heat. Gently sweat the onions for 10–15 minutes until they are just starting to brown. Add the garlic paste and chilli, and cook for another 10–15 minutes, stirring frequently, until everything looks almost caramelized. Tip in the chopped tomatoes, adding a little water if the mixture is too dry; if using canned tomatoes, add the entire contents of the can, juice and all. Season with salt and pepper.

3 Oil a 2-litre (3½-pint) baking dish. Cover the bottom of the dish with half of the baguette slices, arranged tightly together. Sprinkle with a little olive oil, then spread half of the onion and tomato mixture over the top. Layer with half of the sliced tomatoes, and top with half of the basil leaves. Sprinkle over half of the Parmesan. Repeat the layering process, finishing with the remaining Parmesan. Dot with the mozzarella slices, and drizzle the stock over the top.

4 Bake in the middle of the oven for 40–45 minutes until the top is crusty and golden. Remove from the oven, and leave to cool for 5 minutes before serving. Serve with salad, grilled asparagus, and perhaps some lemony grilled chicken.

Pink Accordion (p55)

Works well with
Beefsteak (pp52–63)
Plum (pp64–73)
Standard globe (pp22–35)

Melted feta on tomatoes and spinach

This is a truly scrumptious salad, where hot oil is poured over feta cheese, which in turn melts to drizzle the tomatoes and spinach with moist and oily juices, which then create their own dressing.

Serves 4 as a starter
Prep time: 15 minutes

100g (3½oz) fresh spinach leaves, rinsed and dried

2 large beefsteak tomatoes, about 550g (1¼lb), sliced

12 kalamata olives stored in olive oil, pitted

1 tbsp capers in brine, rinsed, drained, and gently squeezed dry

8 sun-dried tomato halves stored in olive oil, roughly chopped

1 shallot, finely chopped

100g (3½oz) feta cheese

100ml (3½fl oz) olive oil

2 tbsp sherry vinegar or red wine vinegar

freshly ground black pepper

garlic croutons, to serve (optional) (p152)

1 In a large salad bowl, gently toss the spinach, olives, capers, sun-dried tomatoes, and shallot. Layer the tomatoes on top of the salad, and crumble over the feta.

2 Heat the olive oil in a small heavy pan until quite hot, then drizzle carefully over the salad. Finish with the vinegar, and season with a good grinding of black pepper. Serve straight away, topped with the croutons (if using) or accompanied by some garlicky country-style bread.

A word of warning: As you need to pour hot oil over the ingredients, make sure that they are dry before adding the oil, to avoid splattering. Spin-dry the spinach well; once you have added the tomatoes to the salad, you may need to pat them dry with a bit of kitchen paper, before pouring the oil. If you are worried about pouring the oil directly into your salad bowl, use a wok instead.

Stupice (p27)

Works well with
Beefsteak (pp52–63)
Standard globe (p22–35)
Plum (pp64–73)

Pan-fried marrow and green tomatoes

Perfect for when your garden is producing at its peak and you are awash in a glut of fresh vegetables, this recipe uses marrows, but courgettes or pumpkins also work well. Unripe tomatoes provide the perfect complement, while pimentón adds a deliciously rich kick.

Serves 6
Prep time: 20 minutes
Cooking time: 15 minutes

2 tbsp butter
2 tbsp olive oil
2 onions, sliced
1 small marrow, about 1kg (2¼lb), peeled, halved, deseeded, and cut into 1cm (½in) slices
8 unripe tomatoes, about 650–675g (1¼–1½lb), halved and sliced
1 tsp pimentón (Spanish smoked paprika), or to taste
juice of 2 oranges
sea salt and freshly ground black pepper

1 In a frying pan, melt the butter over a medium heat, and sauté the onions for about 10 minutes until they are very soft and caramelized.

2 Add the marrow slices, and cook on both sides for about 10 minutes or until they feel quite soft. If the mixture becomes too dry, add a little extra olive oil. Tip in the tomato slices, and sauté for a further 2–3 minutes. Sprinkle in the pimentón, and season with salt and pepper.

3 To finish, increase the heat under the pan slightly, and add the orange juice, a little at a time, to deglaze the pan. Allow to reduce for a couple of minutes until you have a sticky sauce. Serve hot as a breakfast dish with eggs and bacon, or for an outdoor brunch.

Works well with
Standard globe (pp22–35)
Beefsteak (pp52–63)
Plum (pp64–73)

Tomato salad with tarragon, lemon zest, and capers

People often bemoan the fact that today's tomatoes are tasteless. Well, a good thing about working with something that is arguably not very tasty is that it can act as a sponge for other, stronger flavours. If you do have flavourful tomatoes on hand, that can only be a bonus.

Serves 4
Prep time: 20 minutes

6 ripe tomatoes, about 450g (1lb), skinned, deseeded, and cut into strips

2 tbsp roughly chopped tarragon

2 tbsp capers in brine, drained, rinsed, and finely chopped

2 tsp grated lemon zest

2 tbsp extra virgin olive oil, or more to taste

1 Mix together all the ingredients in a glass or ceramic bowl.

2 Leave to stand for 10 minutes to allow the flavours to infuse, then serve as a side salad. It works particularly well with grilled fish or chicken.

Works well with
Beefsteak (pp52–63)
Plum (pp64–73)

Tomato, peach, and strawberry salad

Fruit salad with a twist! Fresh and zinging with flavour, it is also a visual feast. Even if its fruits come into season at different times in your garden, depending on variety and where you live, you may still be able to catch them all at the same time. If you can't, go shopping.

Serves 4
Prep time: 10 minutes

4–6 ripe tomatoes, about 350–450g (12oz–1lb)

3 ripe peaches

15 strawberries

dash of extra virgin olive oil

dash of balsamic vinegar

handful of fresh mint leaves, shredded

sea salt and freshly ground black pepper

1 Cut the tomatoes in half. Halve and stone the peaches, and cut into long, thin sections. Halve the strawberries, discarding the hulls.

2 Put the tomatoes, peaches, and strawberries in a glass or ceramic serving dish. Add a dash of olive oil and another one of balsamic vinegar. Season with salt and plenty of black pepper, and scatter over the shredded mint. Using your fingertips – clean, of course – gently toss the salad so that the fruit is evenly coated in the dressing. Serve immediately.

Jelly Bean (p40)

Works well with
All tomatoes (pp22–77)

Fresh tomatoes marinated with garlic

Particularly delicious on a hot summer's day, there really could be nothing simpler. The secrets to success are using only the choicest ingredients and allowing time for marinating. The salt draws out moisture from the tomatoes, creating their own tempting dressing.

Serves 4
Prep time: 10 minutes
Marinating time: 2 hours

4 large beefsteak tomatoes, about 1.1kg (2½lb), sliced

1 or 2 garlic cloves

4 tbsp extra virgin olive oil

1–2 tbsp good-quality sherry vinegar

handful of fresh basil leaves, shredded (optional)

sea salt and freshly ground black pepper

1 Put the tomatoes in a shallow glass or ceramic dish.

2 Using a mortar and pestle, pound the garlic to a purée with a little sea salt. Add to the tomatoes with the oil and vinegar. Season generously with a good sprinkling of sea salt and pepper. Gently mix well so that all the ingredients are evenly dispersed. Leave to marinate in a cool part of the kitchen for at least 2 hours.

3 Serve at room temperature – this is important – as an accompaniment to grilled meat or fish, or a side salad for lunch, topped with the shredded basil (if using). Make sure that there is plenty of fresh, crusty bread on the table to soak up the delicious juices.

Variation If you like, use any other fresh herb you fancy to top the marinated salad – try rosemary, oregano, thyme, flat-leaf parsley, or tarragon.

Works well with
Beefsteak (pp52–63)
Plum (pp64–73)
Standard globe (pp22–35)

Dressings

With their intrinsic acidity, tomatoes are a good ingredient for a salad dressing. Next time you are using tomatoes and are about to get rid of the juicy inside with its seeds, think about keeping this and using it as a semi-substitute for vinegar or lemon juice in a dressing.

Makes enough to dress a salad for 4
Prep time: 10 minutes

Proportions for basic dressing

2 tbsp sieved tomato juices (see step 1)

2 tbsp good-quality extra virgin olive oil

2 tsp vinegar such as cider vinegar, apple vinegar, or balsamic vinegar

sea salt and freshly ground black pepper to taste

1 Here is a rough guide to preparing a tomato-based dressing. Adjust the amounts up and down accordingly, to make the amount you need. First, core and deseed a juicy tomato – plum, cherry, and standard globe tomatoes are best for this. Keep the seeds plus any juices, and blend or process until smooth. Pass through a fine sieve if you like.

2 To this mixture, add the same amount by volume of good-quality extra virgin olive oil and a little vinegar. Season with salt and pepper. Whisk with a fork, taste, and adjust the flavours to suit you. And there you have it – a simple yet tasty dressing ready to drizzle over your favourite salad.

Variations

- A little finely chopped fresh dill and some freshly toasted and ground cumin seeds, transforms this dressing – good for roasted tomato, carrot, and beetroot salads.

- Add some finely sliced fresh basil leaves and freshly puréed garlic (use a mortar and pestle to crush the garlic with a little salt) to the basic dressing, and use balsamic vinegar – perfect for summery tomato salads.

- For a citrussy dressing with a Spanish touch, add a pinch or so of pimentón (Spanish smoked paprika), and use freshly squeezed orange or lemon juice, or both, instead of vinegar – ideal for a red pepper, tomato, and orange salad.

Works well with
Plum (pp64–73)
Cherry (pp36–51)
Standard globe (pp22–35)

Turkish shepherd's salad

A beautiful combination of colours and contrasting flavours and textures, this wonderful wintry salad is a definite treat during the chilly months of the year. It can also be made 3–4 hours in advance. Simply prepare the salad, then leave it in a cool place until it's needed.

Serves 6
Prep time: 20 minutes

¼ red cabbage, about 200g (7oz), quartered, cored and very finely sliced

70ml (2½fl oz) extra virgin olive oil

juice of 1 lemon, or to taste

sumac, to taste

5 large ripe tomatoes, about 450g (1lb), cut into wedges and preferably roasted (p156)

2 large roasted beetroots, sliced

½ red onion, very finely chopped

5 tbsp finely chopped flat-leaf parsley

3 tbsp finely chopped dill

5 tbsp fresh mint leaves

sea salt and freshly ground black pepper

1 In a large bowl, mix together the cabbage and olive oil. Add the lemon juice, a little at a time, until you are happy with the taste. Sprinkle the cabbage with a little sumac, and season with salt and pepper.

2 Stir well to make sure that the cabbage is thoroughly covered with dressing, then add the rest of the ingredients. Toss through very well so that all the salad is well oiled and herby. Season generously with pepper and just a little salt – and extra sumac if needed. Serve at room temperature.

Note Sumac is a berry with citrussy tones, which comes from Turkey. It is delicious in salads and with grilled meats. If you have trouble finding it add some lemon juice or red pepper flakes instead.

Floridity (p70)

Works well with
Plum (pp64–73)
Standard globe (pp22–35)

Avocado, tomato, and lime salad

Ripe, unblemished avocados and ripe tomatoes are one of those culinary "marriages" that go together very well indeed. This simple salad focuses on using the choicest ripe ingredients. A fresh vinaigrette invigorated with lime zest makes it simply irresistible.

Serves 4 as a side dish
Prep time: 10 minutes

1 tbsp pumpkin seeds

2 large ripe, unblemished avocados

4 ripe beefsteak tomatoes, about 1–1.1kg (2¼–2½lb), finely sliced

For the lime dressing

2 tbsp extra virgin olive oil

1 tbsp white wine vinegar

1 tsp tequila (optional)

zest and juice of ½ lime

sea salt and freshly ground black pepper

1 In a dry frying pan over a medium heat, toast the pumpkin seeds for a couple of minutes until they start to pop, taking care not to scorch them. Tip into a small bowl, and set aside.

2 To make the dressing, whisk together the oil, vinegar, tequila (if using), lime zest, and a squirt of lime juice in a small bowl. Season with salt and pepper. Taste, and adjust the seasoning if needed.

3 Halve, peel, and stone the avocados, and cut into slices lengthways. Arrange alternating layers of avocado and tomato slices on a serving dish or platter, and pour over the vinaigrette. Scatter over the reserved toasted pumpkin seeds, and serve immediately.

Works well with
Beefsteak (pp52–63)
Cherry (pp36–51)

Caponata

One of Sicily's signature dishes, caponata's combination of sweet and sour flavours is typical of the island's distinctive cuisine. Great served with fresh crusty bread as an antipasto, it also makes an excellent side dish. As a bonus, it keeps well in the refrigerator for a few days.

Serves 4 as a side dish
Prep time: 15 minutes
Cooking time: 25–30 minutes

100ml (3½fl oz) olive oil

3 aubergines, cut into 2cm (1in) cubes

2 onions, sliced

100g (3½oz) celeriac or celery, finely chopped

100g (3½oz) green olives in brine, pitted

50g (1¾oz) capers in brine, drained, rinsed, and gently squeezed dry

1 x 400g (14oz) can chopped plum tomatoes, drained, or 5 –6 ripe plum tomatoes, about 450–550g (1–1½lb), skinned

50g (1¾oz) sugar

60ml (2fl oz) cider vinegar or white wine vinegar

¼ tsp cocoa powder or dark chocolate (at least 70% cocoa solids), finely grated

sea salt and freshly ground black pepper

freshly chopped flat-leaf parsley, to serve

1 Heat 75ml (2½fl oz) of the olive oil in a heavy frying pan over a high heat. Add the aubergine, and quickly fry for 8–10 minutes until starting to turn golden brown. Remove from the pan with a slotted spoon, and leave to drain on kitchen paper.

2 In the same pan, heat the remaining oil, and sauté the onions for a further 8–10 minutes until golden brown.

3 Meanwhile, blanch the celeriac or celery in a small pan of rapidly boiling water for 1 minute. When the celeriac or celery is nearly cooked, tip into a colander or bowl, and refresh with cold water. Drain.

4 Add the celeriac or celery with the olives and capers to the sautéed onions, then add the tomatoes (or fresh tomatoes, if using), sugar, vinegar, and a tiny amount of chocolate. Cook for 8–10 minutes over a medium heat, stirring with a wooden spoon from time to time. Remove from the heat, add the reserved aubergines, and season with salt and pepper; be careful with the amount of salt you use, as the olives and capers will already be adding their own saltiness.

5 Transfer to a serving dish, and decorate with some chopped parsley. Serve at room temperature.

Note If you have made the caponata ahead and stored it in the refrigerator until needed, it is important to allow it to come to room temperature before serving.

Works well with
Plum (pp64–73)

Tomato and mascarpone ice cream with raspberries

The flavour of the tomatoes here may be subtle, but it is an integral and entirely delicious element of this rich and creamy concoction. Confirmation, if it were needed, that the tomato is indeed a fruit.

Makes 1 litre (1¾ pints)
Prep time: 30 minutes
Freezing time: 2 hours

250g (9oz) ripe tomatoes, skinned, deseeded, and finely diced (soft-tasting varieties such as vine-ripened cherry, or yellow tomatoes work best)
70g (2½oz) fresh raspberries
1 tbsp sugar
2 tbsp Amaretto liqueur

For the custard base
3 egg yolks
100g (3½oz) caster sugar
2 tbsp runny honey
360ml (12fl oz) single cream
100ml (3½fl oz) double cream
1 tsp almond extract
250g (9oz) mascarpone cheese

1 Put the tomatoes and raspberries in a glass bowl. Add the sugar and Amaretto. Chill until needed.

2 Whisk together the egg yolks, caster sugar, and honey in a bowl until smooth, pale, and thick.

3 To make the custard, gently heat the single cream, double cream, and almond extract to just below boiling point. Add a little of the hot cream mixture to the egg yolk mixture, and whisk quickly to incorporate. Add a little more hot cream, and whisk again. Pour this mixture back into the pan, and stir constantly with a wooden spoon until the custard coats the back of the spoon. Remove from the heat. Whisk in the mascarpone until smooth, then chill until cold.

4 Churn the custard in an ice-cream machine for about 20 minutes or according to the manufacturer's instructions, then spoon into a freezerproof container. Carefully but thoroughly stir through the tomato and raspberry mixture, and freeze until firm. If you don't have a machine, put the custard and mascarpone mixture in a container, and freeze for 1 hour. Whisk well to break up any ice crystals, then repeat this process twice more. Incorporate the fruits, then freeze until firm. Remove from the freezer 15 minutes before needed, to allow it to soften a little.

Tommy Toe (p38)

Works well with
Cherry (pp36–51)
Standard globe (pp22–35)

Tomato and lemon marmalade

This simple marmalade can readily be made with red tomatoes, but use either yellow or brown ones if you have them. It will keep for up to a year if stored in a cool, dark place. Keep in the refrigerator after opening.

Makes about 1.5kg (3lb 3oz)
Prep time: 30 minutes
Cooking time: 1½ hours

5 large unwaxed lemons,
  zest peeled and cut into long,
  thin strips
1kg (2¼lb) tomatoes, skinned,
  cored, and cut into quarters
1kg (2¼lb) granulated sugar

1 Put the lemon strips in a small pan with a little water to cover. Simmer for 15 minutes or until soft. Remove from the heat, and set aside.

2 Cut the peeled lemons in half, and squeeze out all the juice, keeping any pips. Set the juice aside. Scoop out any remaining flesh, and put in a glass or stainless-steel bowl with the reserved pips. Scoop out the seeds and juices from the tomato quarters, and put in the same bowl; mix well. Tip onto some clean muslin, and tie tightly – this is your pectin bag.

3 Cut the tomato quarters into slices, mix with the lemon zest (and its cooking liquid), and tip into a large stainless-steel pan. Measure the reserved lemon juice, and make up to 1.8 litres (3¼ pints) with cold water. Add this liquid to the pan, along with the pectin bag.

4 Bring the mixture to the boil, reduce the heat, and simmer for about 45 minutes, stirring occasionally, until it thickens and the zest is soft. Remove the bag, squeezing all the juice back into the pan. Add the sugar, stir well, and simmer gently for 20–30 minutes until setting point has been reached.

5 Pot into hot, sterilized jars with tight-fitting lids, seal straight away, and label (pp142–3).

Testing for setting point
Turn off the heat, and drop a teaspoon of the marmalade onto a clean cold saucer. Let it cool, then push your finger into it – if the marmalade wrinkles, setting point has been reached.

Brown Berry (p45)

Works well with
Cherry (pp36–51)
Standard globe (pp22–35)

Glossary

abscission layer – a zone of cells whose breakdown causes separation of a leaf or fruit from the stem

annual – a plant that completes its lifecycle in one growing season

bush – a plant that produces a number of sideshoots

calyx – the collective term for the green sepals of the flower that protect it in the bud stage and form a spider-like structure on top of the ripe fruit

chlorophyll – the green plant pigment mainly responsible for light absorption

cordon – a plant generally restricted to one main stem

cotyledon – the first leaf or leaves to emerge after germination of a seed

cutworm – the larvae of various noctuid moths

cross-pollination – the transfer of pollen from the anther of a flower on one plant to the stigma of a flower on another plant

determinate plant – a bushy or dwarf tomato plant

dwarf – a plant that contains a dwarfing gene, making it very compact

F1 hybrid – the term stands for 'First filial generation' and refers to a cross of two pure breeding parental lines

fungus gnat – midge-like flies, up to 4mm

genus – a category in plant classification between family and species

hybrid – a plant resulting from a cross between two distinct parents

heirloom – an old/treasured open-pollinated plant

indeterminate plant – tall or cordon plants that can grow to an indefinite length

knuckle – the point were the calyx of the tomato fruit joins the stem

leading shoot – the main, usually central, stem of a plant

leaf axil – the upper angle between a leaf and a stem

locule – a cavity or chamber within the fruit

loam – a term usually used imprecisely to denote a rich soil with a balanced mix of clay, sand, and humus

module – individual containers used in multiples for sowing seeds

nematode – a worm-like animal also call an eelworm

open pollinated – seed produced from natural pollination, which can result in varied plants, although, as most tomatoes are self-fertile their offspring tend to be consistent

perennial – any plant living for at least three growing seasons

Useful resources

photosynthesis – the process by which plants use sunlight to convert carbon dioxide and water into carbohydrates

pollen – male sex cells produced by stamens

pollination – the transfer of pollen from anthers to stigmas

potash – any of several compounds containing potassium

propagator – a structure providing a humid atmosphere for raising seedlings

seed modules – individual containers used in multiples for sowing seeds

semi-determinate plant – term usually used for cordons that grow to only 1–1.2m (3–4ft) or for those intermediate in growth habit between bush and cordon

sideshoot – a stem that arises from the side of a main shoot

Solanaceae – the plant family to which tomatoes and potatoes belong

species – a category in plant classification containing very similar individuals

stigma – the part of the female sex organ that receives pollen

truss – a compact cluster of flowers or fruits

variety – a grouping of plants having distinctive features that persist through successive generations

Gardens to visit

Audley End, Saffron Walden, UK www. gardenorganic.org.uk/gardens/audley.php

West Dean Gardens, Chichester, UK www.westdean.org.uk

Heronswood at Dromana, and St Erth at Blackwood are The Diggers Club gardens www.diggers.com.au

Societies

Association Kokopelli, France www.kokopelli-seeds.com

Garden Organic Ryton, Coventry, UK (formerly the Henry Doubleday Research Association (HDRA)) www.gardenorganic.org.uk

Seed Savers' Network, Australia www.seedsavers.net

Seed Suppliers

Simpsons Seeds, Warminster, UK www.simpsonsseeds.co.uk

Thompson and Morgan, UK www.thompson-morgan.com

Eden Seeds/Select Organic, Australia www.edenseeds.com.au/ www.selectorganic.com.au

Green Harvest, Australia www.greenharvest.com.au

Cookery

www.sofiacraxton.co.uk

Index

Acknowledgments

Dorling Kindersley would like to thank: Siobhan O'Connor for her excellent recipe editing; Annelise Evans for all her editorial help; Carolyn Hewitson and Kenny Grant for their design assistance; Graham Rae, William Reavell, Howard Rice, and Sarah Ashun for the excellent photography; Jo Walton, Romaine Werblow, and Karen Forsythe for their picture research; Katie Giovanni, the food stylist, for her expertise; Sue Rowlands, the prop stylist; Jim Buckland, Sarah Wain, and Shirley Tasker at West Dean Gardens, Sussex, for allowing us to photograph their wonderful tomato collection; Jane, Amanda, and Lesley at Not Just Food for their efficient recipe testing; Hilary Bird for creating the index; and the following for their kind donation of tomatoes: Jane and Matt Simpson at Simpson's Seeds, Kieran Devine at Wight Salads Group; Jim Arbury at Wisley Gardens; Sharon MacGregor at De Ruiter Seeds; Mike Thurlow at Audley End; Terry Marshall; Pennard Plants; Bernard Sparkes at Melrow Salads; and Jessica Tsang.

Gail Harland would like to thank: Andrew Roff at Dorling Kindersley for editing with great good humour and photographers Graham Rae and the multi-talented Howard Rice. Many people at seed firms, particularly Caroline Rush at Thompson and Morgan. Grateful thanks also go to my sons Ashley and Jonathan who have been chief tasters of a range of sometimes strange tomatoes over the years. Somehow they still seem to believe that tomatoes should be round and red.

Sofia Larrinua-Craxton would like to thank: Savi Sperl for her endless enthusiasm; Barbara and Gaetano Mansi, for sharing the Ragu recipe; Cecilia Gutierrez Elizondo, for making that memorable gazpacho, which we loved so much; Antonella Tagliapietra and Philip Oakes for their hospitality at Arte Culinaria and for the recipe for ox cheeks; New Covent Garden Food Co. for giving me the chance to do so much work in the UK and USA, thanks also go to you for the Tomato Borscht recipe; Eric, Rosie, Sally, and everyone at Books for Cooks – you are wonderful!; MC, Andrew, and everyone at DK; to all my family, including Merida who wanted her name printed "in my next book"; and last but not least I'd like to thank my husband Oliver to whom I would like to dedicate my share of this book.

Picture credits The publisher would like to thank the following for their kind permission to reproduce their photographs: (Key: b-below/bottom; c-centre; l-left; r-right; t-top) Alamy Images: Arco Images GmbH 87; Nigel Cattlin 115bl, 115br; Company 11bl; John Glover 11tl; Stan Kujawa 115tr; Gerald Majumdar 113; Helene Rogers 86b; W Atlee Burpee & Co.: 17cr, 32t; De Ruiter Seeds: 47b; DK Images: Will Heap 2; Barrie Watts Collection 85br; FLPA: Nigel Cattlin 115tl; The Garden Collection: Torie Chugg 103tl; Liz Eddison/Design: Phillipa Pearson 85; Jane Sebire 105bl; Neil Sutherland 121; Getty Images: 105tl; Gail Harland: 28b; Marshalls Seeds: 17tr, 58tl; Photolibrary: 78-79; Pro-Veg Seeds Ltd: 47c; Sallie Sprague: 34c; Diana Stek: 21cl, 25b, 77c; Suttons Seeds: 32b, 83t; Constance Toops: 29t, 35tr, 53t, 55b, 58c, 74t, 77b, 168b; www.crocus.co.uk: 97All other images © Dorling Kindersley. For further information see: www.dkimages.com